Lecture Notes in Artificial Intelligence   13119

Subseries of Lecture Notes in Computer Science

More information about this subseries at https://link.springer.com/bookseries/1244

Richard Chbeir · Yannis Manolopoulos ·
Rajendra Prasath (Eds.)

Mining Intelligence and Knowledge Exploration

9th International Conference, MIKE 2021
Hammamet, Tunisia, November 1–3, 2021
Proceedings

Springer

Editors
Richard Chbeir (iD)
Université de Pau et des Pays de l'Adour
Anglet, France

Yannis Manolopoulos (iD)
Open University of Cyprus
Nicosia, Cyprus

Rajendra Prasath (iD)
Indian Institute of Information Technology
Sri City, Chittoor
Andhra Pradesh, India

ISSN 0302-9743   ISSN 1611-3349 (electronic)
Lecture Notes in Artificial Intelligence
ISBN 978-3-031-21516-2   ISBN 978-3-031-21517-9 (eBook)
https://doi.org/10.1007/978-3-031-21517-9

LNCS Sublibrary: SL7 – Artificial Intelligence

This Springer imprint is published by the registered company Springer Nature Switzerland AG
The registered company address is: Gewerbestrasse 11, 6330 Cham, Switzerland

Preface

This volume contains the revised collection of papers presented at MIKE 2021: the 8th International Conference on Mining Intelligence and Knowledge Exploration held during November 1–3, 2021, as a virtual event hosted from Hammamet, Tunisia (http://www.mike.org.in/2021/). MIKE 2021 received 61 qualified submissions from authors in 12 countries and each qualified submission was reviewed by a minimum of two Technical Program Committee members using the criteria of relevance, originality, technical quality, and presentation. This rigorous XX blind review process finally resulted in 16 of those submissions being accepted for presentation at the conference. Hence, the overall acceptance rate for this edition of MIKE is 26.23%.

The International Conference on Mining Intelligence and Knowledge Exploration (MIKE) is an initiative focusing on research and applications on various topics of human intelligence mining and knowledge discovery. Human intelligence has evolved steadily over several generations, and today human expertise is excelling in multiple domains and in knowledge-acquiring artifacts. The primary goal of MIKE 2021 was to focus on the frontiers of human intelligence mining toward building a body of knowledge in this key domain. The focus was also to present state-of-the-art scientific results, to disseminate modern technologies, and to promote collaborative research in mining intelligence and knowledge exploration. At MIKE 2021, specific focus was placed on the "Learning to innovate using Internet-of-Things".

MIKE 2021 featured 10 tracks, each led by 2–3 track coordinators who handled submissions falling in their areas of interest. The involvement from each of them along with the supervision of the Program Chairs ensured the selection of quality papers for the conference. The track coordinators took responsibility for the tasks assigned to them after we started circulating the first call for papers. This is reflected in the impact and quality of every paper appearing in the proceedings.

The accepted papers were chosen on the basis of research excellence, which provides a body of literature for researchers involved in exploring, developing, and validating learning algorithms and knowledge-discovery techniques. Accepted papers were grouped into various subtopics including evolutionary computation, knowledge exploration in IoT, artificial intelligence, machine learning, data mining and information retrieval, medical image analysis, pattern recognition and computer vision, speech/signal processing, text mining and natural language processing, intelligent security systems, smart and intelligent systems, and other areas not included in the above list. Researchers presented their work and had an excellent opportunity to interact with eminent professors and scholars in their area of research. All participants benefited from discussions that facilitated the emergence of new ideas and approaches.

We were pleased to have notable scholars serving as advisory committee members for MIKE 2021, who are listed in the following pages. We sincerely express our gratitude to Bayya Yegnanarayana, International Institute of Information Technology, Hyderabad, India, and Chaman Lal Sabharwal, Missouri University of Science and Technology,

USA. Their guidance, suggestions, and constant support, were invaluable in planning various activities for MIKE 2021.

Several eminent scholars, including Sankar Kumar Pal, Indian Statistical Institute, Kolkata; Sung-Bae Cho, Yonsei University, South Korea; Alexander Gelbukh, Instituto Politecnico Nacional, Mexico; and N. Subba Reddy, Gyeongsang National University, South Korea, also extended their kind support in guiding us to organize a MIKE conference that was even better than the previous edition.

Payam Barnaghi from Imperial College London, UK, delivered the keynote address on "Developing a Digital Platform for Remote Healthcare Monitoring"; Rajiv Ratn Shah from the Indraprastha Institute of Information Technology, Delhi, India, delivered an invited talk on "Leveraging Multimodal Data and AI for Social Good"; Georgios N Yannakakis from the University of Malta, Malta, delivered an invited talk on "Big data and small data: the challenge is in the interpretation"; and finally Johann Gamper from the University of Bozen-Bolzano, Italy, delivered an invited talk on "Key Challenges in Processing Temporal and Time Series Data".

A large number of eminent professors, well-known scholars, industry leaders, and young researchers participated in making MIKE 2021 a great success. We recognize and appreciate the hard work put in by each individual author of the articles being published in these proceedings. We also express our sincere thanks to the National Institute of Technology, Goa, India, for hosting MIKE 2021.

We thank the Technical Program Committee members and all reviewers for their timely and thorough participation in the reviewing process. We appreciate the time and effort put in by the members of the local organizing team at the University of Jendouba, Tunisia, and IIIT Sri City, India. We are very grateful to all our sponsors for their generous support to MIKE 2021.

Finally, we acknowledge the support of EasyChair in the submission and review processes.

We are also very pleased to express our sincere thanks to the team at Springer their support in publishing the proceedings of MIKE 2021.

November 2021

Richard Chbeir
Yannis Manolopoulos
Rajendra Prasath

Organization

General Chairs

Richard Chbeir   University of Pau and Pays of Adour, France
Rajendra Prasath   Indian Institute of Information Technology Sri
                                    City, India

Advisory Committee

Adrian Groza   Technical University of Cluj-Napoca, Romania
Agnar Aamodt   Norwegian University of Science and Technology,
                                    Norway
Aidan Duane   Waterford Institute of Technology, Ireland
Alexander Gelbukh   Instituto Politécnico Nacional (IPN), Mexico
Amit A. Nanavati   IBM Research, New Delhi, India
Anil Vuppala   IIIT Hyderabad, India
Ashish Ghosh   Indian Statistical Institute, Kolkata, India
B. Yegnanarayana   IIIT Hyderabad, India
Bjorn Gamback   NTNU, Norway
Chaman Lal Sabharwal   Missouri University of Science and Technology,
                                    USA
Debi Prosad Dogra   Indian Institute of Technology, Bhubaneswar,
                                    India
Genoveva Vargas-Solar   CNRS, France
Grigori Sidorov   CIC - IPN, Mexico
Hrishikesh Venkataraman   IIIT Sri City, India
Ildar Batyrshin   Instituto Politécnico Nacional (IPN), Mexico
Kazi Shah Nawaz Ripon   NTNU, Norway
Krishnaiyya Jallu   BHEL, India
Mandar Mitra   Indian Statistical Institute, Kolkata, India
Manish Shrivastava   IIIT Hyderabad, India
Maunendra S. Desarkar   Indian Institute of Technology, Hyderabad, India
N. Subba Reddy   Gyeongsang National University, South Korea
Niloy Ganguly   Indian Institute of Technology, Kharagpur, India
Nirmalie Wiratunga   Robert Gordon University, Scotland
P.V. Rajkumar   Texas Southern University, USA
Paolo Rosso   Universitat Politecnica de Valancia, Spain
Philip O'Reilly   University College Cork, Ireland

Pinar Ozturk	NTNU, Norway
Radu Grosu	TU Wien, Austria
Rajarshi Pal	IDRBT, India
Ramon Lopez de Mantaras	IIIA - CSIC, Spain
Saurav Karmakar	GreyKarma Technologies India
Sudeshna Sarkar	Indian Institute of Technology, Kharagpur, India
Sudip Misra	Indian Institute of Technology, Kharagpur, India
Susmita Ghosh	Jadavpur University, India
T.Kathirvalavakumar	VHNSN College (Autonomous), India
Tanmoy Chakraborty	IIIT Delhi, India
Tapio Saramäki	Tampere University of Technology, Finland
V. Ravi	IDRBT, India
Vasile Rus	University of Memphis, USA
Vasudeva Verma	IIIT Hyderabad, India
Yannis Stylianou	University of Crete, Greece

Technical Program Committee

Adam Krzyzak	Concordia University, Canada
Alexander Ryjov	Moscow State University, Russia
Anca Hangan	Technical University of Cluj-Napoca, Romania
Anca Marginean	Technical University of Cluj-Napoca, Romania
Antonino Staiano	Università di Napoli Parthenope, Italy
Antonio Morán	University of Leon, Spain
Athanasios Tsadiras	Aristotle University of Thessaloniki, Greece
Birjodh Tiwana	LinkedIn, USA
Bogdan Iancu	Technical University of Cluj-Napoca, Romania
Chaman Lal Sabharwal	Missouri University of Science and Technology, USA
Christos Georgiadis	University of Macedonia, Greece
Ciprian Oprisa	Technical University of Cluj-Napoca, Romania
Costas Iliopoulos	King's College London, UK
Costin Badica	University of Craiova, Romania
Debasis Ganguly	Dublin City University, Dublin, Ireland
Denis Trcek	University of Ljubljana, Slovenia
Dileep A. D.	Indian Institute of Technology, Mandi, India
Elio Mansour	Université de Pau et des Pays de l'Adour, France
Epaminondas Kapetanios	University of Hertfordshire, UK
Eva Onaindia	Polytechnic University of Valencia, Spain
Farah Bouakrif	University of Jijel, Algeria
Florin Leon	"Gheorghe Asachi" Technical University of Iași, Romania
Frantisek Capkovic	Slovak Academy of Sciences, Slovakia

George Tsekouras	University of the Aegean, Greece
Giannis Tzimas	University of Peloponnese, Greece
Giorgio Gnecco	IMT - School for Advanced Studies Lucca, Italy
Gloria Inés Alvarez	Pontificia Universidad Javeriana Cali, Columbia
Goutham Reddy A.	National Institute of Technology, Andhra Pradesh, India
Hans Moen	University of Turku, Finland
Ilias Sakellariou	University of Macedonia, Greece
Ioannis Chamodrakas	National and Kapodistrian University of Athens, Greece
Ioannis Hatzilygeroudis	University of Patras, Greece
Ioannis Karydis	Ionian University, Greece
Isidoros Perikos	University of Patras, Greece
Isis Bonet Cruz	EIA University, Columbia
Jacek Kabziński	Technical University of Lodz, Poland
Jimmy Jose	National Institute of Technology, Calicut, India
Jose Maria Luna	University of Córdoba, Spain
Juan Recio-Garcia	Universidad Complutense de Madrid, Spain
Karam Bou Chaaya	University of Pau and Pays of Adour, France
Kazi Shah Nawaz Ripon	Norwegian University of Science and Technology, Norway
Khouloud Salameh	American University of Ras Al Khaimah, United Arab Emirates
Konstantinos Margaritis	University of Macedonia, Greece
Kostas Karpouzis	Panteion University, Greece
Lasker Ershad Ali	Peking University, China
Maciej Ogrodniczuk	Polish Academy of Sciences, Poland
Manolis Maragoudakis	University of the Aegean, Greece
Maristella Matera	Politecnico di Milano, Italy
Martin Holena	Czech Academy of Sciences, Czech Republic
Mihaela Oprea	Petroleum-Gas University of Ploiesti, Romania
Mikko Kolehmainen	University of Eastern Finland, Finland
Mirjana Ivanovic	University of Novi Sad, Serbia
Muhammad Khurram Khan	King Saud University, Saudi Arabia
N. S. Reddy	Gyeongsang National University, South Korea
Nagesh Bhattu	National Institute of Technology, Andhra Pradesh, India
Nikolaos Polatidis	University of Brighton, UK
Oana Iova	National Institute of Applied Sciences, Lyon, France
Odelu Vanga	Indian Institute of Information Technology Sri City, India
Ondrej Krejcar	University of Hradec Kralove, Czech Republic

Additional Reviewers

Christopher, Gladis
Hammemi, Hamza
Haube, Giovanni
Kathirvalavakumar, T
Kumaran, T
Mignone, Paolo
Zuluaga, Jhony Heriberto Giraldo

Contents

Type 2 Diabetes Prediction from the Weighted Data

A. Suriya Priyanka[1], T. Kathirvalavakumar[2(✉)], and Rajendra Prasath[3]

[1] Department of Information Technology, V.H.N. Senthikumara Nadar College, Madurai Kamaraj University, Virudhunagar, Tamil Nadu 626001, India
suriyapriyanka@vhnsnc.edu.in
[2] Research Centre in Computer Science, V.H.N. Senthikumara Nadar College, Madurai Kamaraj University, Virudhunagar, Tamil Nadu 626001, India
kathirvalavakumar@yahoo.com
[3] Computer Science and Engineering Group, Indian Institute of Information Technology, Sri City, Chittoor, Andhra Pradesh 517646, India
rajendra.prasath@iiits.in

Abstract. The world is struggled also with diabetes as the number of diabetic patients is increased rapidly. As the diabetes lead to create related critical diseases, more researchers are interested to predict the diabetes. This paper proposes a model for predicting the type 2 diabetes using KNN with the weighted trained data of the identified significant features. The weighted data is calculated from the probability of occurrence of a data in a neighborhood. Pima Indian Diabetes dataset is used in the experiment. The obtained result show that proposed method gives better result than the existing results in the literature.

Keywords: Prediction · Type 2 diabetes · Network pruning · KNN · Weighted data

1 Introduction

Now-a-days everyone in the world is familiar with the word "Diabetes". Both developed and developing countries are facing challenges to prevent, predict and cure diabetes. Diabetes, also called as diabetes mellitus is a set of metabolic disorders identified by high blood glucose levels over a prolonged period [1]. According to a report of the International Diabetes Federation in 2017, there were 425 million diabetics in the world at the time, and it was also concluded that the number will increase to 625 million by 2045 [2]. The people affected by diabetes suffer with high levels of glucose causing diseases like heart diseases, kidney diseases, cardiovascular diseases and it leads to serious problems for pregnant women [3]. So it is mandatory for early diagnosis of diabetes to prevent from side effects. Since there is no such medicines exist to cure diabetes, it is necessary to control and prevent diabetes. Now-a-days researchers are focusing to develop model for predicting diabetes using Machine Learning and Deep Learning techniques.

© Springer Nature Switzerland AG 2022
R. Chbeir et al. (Eds.): MIKE 2021, LNAI 13119, pp. 1–12, 2022.
https://doi.org/10.1007/978-3-031-21517-9_1

Alehegn et al., [4] have developed Proposed Ensemble Method (PEM). They have used SVM, Naivet Net, Decision Stump classification algorithm and combined the prediction of them into one to increase the prediction accuracy of the algorithm. Hasan et al., [5] have proposed weighted ensembling of different Machine Learning (ML) models and estimated the weights from the corresponding Area Under ROC Curve (AUC). The model has used Pima Indian Diabetes dataset. Outlier rejection and feature selection is used to improve the performance of the classifiers. Various ML classifiers (KNN, Decision Tree, Random Forest, AdaBoost, Naive Bayes and XGBoost), Multilayer Perceptron and their proposed model are compared and found that the proposed ensemble model (AdaBoost + XGBoost) produced better performance.

Zou et al., [6] have applied Decision Tree, Random Forest and Neural Network to predict diabetes mellitus on physical examination data in Luzhou, China and applied the same to existing Pima Indian diabetes dataset. They have used PCA and mRMR to reduce the dimensionality of the dataset. The classification result is compared over the dataset with extracted features and dataset with all features. Better result is obtained from the Random Forest when all features were used. Anwar et al., [7] have surveyed the Artificial Intelligence (neural networks, machine learning, deep learning, hybrid methods and/or stacked-integrated use of different machine learning algorithms) approaches for diabetes prediction on Pima Indians diabetes dataset and concluded that deep learning combined with other algorithms give better results. Zhou et al., [2] have developed deep learning model for predicting diabetes on Pima Indians diabetes dataset and diabetes type dataset (Data World). The model was mainly built using deep neural network in the hidden layers of the network. Dropout regularization is used to prevent overfitting and used binary cross-entropy loss function to improve the effectiveness and adequacy of the proposed model. Bukhari et al., [8] have developed an improved ANN model using an artificial backpropagation scaled conjugate gradient neural network (ABP-SCGNN) algorithm to predict diabetes on the PID dataset.

Sneha et al., [9] have proposed optimal feature selection algorithm, based on the sensitivity of the dataset and applied various ML algorithms for the prediction of diabetes mellitus. Various models namely decision tree, Random forest, SVM and Naive Bayes are compared but obtained better result from decision tree and Random Forest. Kopitar et al., [10] have used Linear Regression, Regularised generalized linear model, Random Forests, Extreme Gradient Boosting (XGBoost) and Light Gradient Boosting for diabetes prediction on routinely collected data from EHR available in ten healthcare centers of Slovenia. The higher stability of selected variables over time contributes to simpler interpretation of the models and concluded that interpretability and model calibration should also be considered in development of clinical prediction models. Muhammad et al., [11] have applied predictive supervised machine learning models (logistic regression, SVM, KNN, Naïve Bayes, Random Forest and gradient boosting) on the diabetes dataset collected from Murtala Mohammed specialist hospital, Kano. They have found that Random Forest produced bettter result in terms of accuracy on diabetes prediction. Maniruzzaman et al., [12] have developed ML based model using Naïve Bayes, Decision tree, AdaBoost and Random Forest to predict diabetes. The model has used diabetes dataset, conducted in 2009–2012, derived from the National Health and Nutrition Examination Survey. The comparative analysis found that Random Forest produced better result than other models.

Gupta et al., [13] have implemented Extra Trees classifier method for feature selection and used KNN classification model for diabetes prediction. The model classified the samples of PID dataset. Le et al., [14] have applied a wrapper-based ML model to predict the early onset of diabetes patients. They have used IQR (Interquartile Range) method for outlier detection. Then they have applied Grey Wolf Optimize (GWO) and Adaptive Particle-Grey Wolf optimization (APGWO) for feature selection to enhance the performance of Multilayer Perceptron (MLP).

Ahmed et al., [15] have built diabetes predictive models for the Electronic Health Records of five different Saudi hospitals using Logistic Regression, SVM, Decision Tree and two ensemble learners (Random Forest and Ensemble Majority Voting). 10-fold cross-validation is used for model evaluation. Insignificant features are eliminated by feature permutation and hierarchical clustering.

In this paper, type 2 diabetes prediction model is proposed for the Pima Indians Diabetes (PID) dataset. Data preprocessing is an essential part for improving the performance of any model. The missing values in each feature of the dataset except number of times pregnant feature are replaced with a mean value of the corresponding feature of the patients having the same diagnosed diabetes and normalize the data in each feature. N2PS Pruning algorithm[16] is used in the artificial neural network for feature selection. Probability is used to find the weighted value for all the values under the extracted features. KNN is used for predicting type 2 diabetics from the weighted data. Rest of the paper is organized as follows: Sect. 2 describes materials and methods, Sect. 3 shows the obtained better experimental results and Sect. 4 concludes the proposed work.

2 Materials and Methods

The PID dataset is collected from UCI machine learning library for the experiment. The dataset contains the attributes namely number of times pregnant, glucose, blood pressure, skin thickness, insulin, body mass index, diabetes pedigree function and age with its diagnosed attribute. It contains totally 768 records with missing values in some of the features. Table 1 describes the features of the dataset.

Data preprocessing is an essential part for improving the performance of any model. In this proposed work, missing values of the features corresponding to diabetic patients and non-diabetic patients are replaced separately. Mean value of each feature is calculated corresponding to diabetic and non-diabetic patients separately and are used for replacing missing values of the respective feature. The value under each feature is normalized by dividing it with the maximum value of the feature.

Single hidden layer feedforward neural network is used for identifying the significant features in the dataset to predict diabetes patient. The neural network is trained with standard backpropagation algorithm. The input layer neurons represent the features of the dataset and are pruned using N2PS pruning algorithm after the network is trained. This procedure extracts the significant attributes for identifying the diabetes. The literature survey says that weighted KNN classification algorithm works well than normal KNN

[17, 18]. In the proposed work, instead of applying weighted KNN, the values in the dataset are converted into weighted value based on probability of the occurrences of each value in its neighborhood. KNN classification algorithm is applied on the calculated weighted value for predicting the type 2 diabetes. The experiment is carried out with 5 fold validation for 5 times. The average result is specified as a result of the experiment.

Table 1. Dataset description

Feature	Description of feature
Pregnancies	Number of times pregnant
Glucose	Plasma glucose concentration in a ah oral glucose tolerance test
Blood pressure	Diastolic blood pressure (mm Hg)
Skin thickness	Triceps skin fold thickness(mm)
Insulin	2 h serum insulin (mu U/ml)
BMI	Body mass index (weight in kg/(height in m)^2)
Diabetes pedigree function	Diabetes pedigree function
Age	Age
Outcome	Class (0 or 1)

2.1 Missing Values Replacement

In the dataset, number of times pregnant, glucose, blood pressure, skin thickness, insulin and BMI features are having missing values and the missing values are represented as zero. One of the features of the dataset is number of times pregnant. It is possible that number of times pregnant may be zero. So this feature is not considered for the replacement procedure. For other features, mean value of the diabetic and non-diabetic patients are calculated separately and are used to replace the missing values in the corresponding feature of patient category.

2.2 Normalization

The values under each feature are normalized by dividing the value with the maximum value of the corresponding feature.

2.3 Feature Extraction

Single hidden layer neural network is considered for identifying the insignificant features. The neural network is trained with the standard backpropagation algorithm. The training is carried out for fixed number of epochs. N2PS pruning algorithm [16] is applied on the trained neural network for extracting the significant features.

2.4 N2PS

In the single hidden layer feedforward neural network, the output of each neuron in a layer is connected with all neurons in the adjacent layer. The output of each neuron is passed to all adjacent layer neurons through its weights. For every input neuron, sum of all input values of the dataset are calculated. The sum of weighted value for the ith input neuron to all hidden layer neuron (s_i) is calculated as follows.

$$si = \sum_{j=1}^{m1} \left| f(tx_{ip}) + w_{ij} \right| \tag{1}$$

where f is a linear function as $f(x) = x$ and $tx_{ip} = \sum_{p=1}^{np} x_{ip}$, here **i** represents input neuron, **m1** represents number of hidden neurons, **np** represents number of patterns in the dataset, x_{ip} represents input value for the ith neuron of the pth pattern, w_{ij} represents the weights corresponding to the ith input neuron passed to j^{th} hidden neuron.

The significance of the input neuron is decided by the following formula.

$$n_i = \begin{cases} \text{insiginificant if } s_i <= \alpha, \alpha = \sum_{i=0}^{m0} s_i/m0 \\ \text{significant} \quad \text{otherwise} \end{cases} \tag{2}$$

where n_i represents i^{th} input neuron, **m0** represents number of input neurons and α is used as a threshold to identify the significance of the neuron. The threshold is the mean of the s_i. If the sum of input values passed from i^{th} input neuron is not greater than the threshold, then corresponding input neuron is considered as insignificant and is pruned.

2.5 Weighted Value

The values under each extracted significant features are sorted. The values under each feature are divided into 5 groups based on its risk factor. Probability for each value under each significant feature is calculated among the neighbors within the group by the following formula. Here neighbors are based on sorted index and not based on distance.

$$Px_i = \frac{Number\ of\ neighbors\ of\ x_i\ in\ the\ group\ belonging\ to\ same\ diagnosis\ attribute\ of\ x_i}{Size\ of\ neighbors} \tag{3}$$

$$Weight\ of\ x_i = x_i * Px_i \tag{4}$$

where x_i represents the ith value under a feature and Px_i represents the probability of ith value of the feature.

2.6 Prediction Model

K-nearest neighbor algorithm is used to predict the type 2 diabetes from the weighted value of the dataset. Euclidean distance measure is used to find the distance. 80% of the weighted dataset (training) is used in the KNN algorithm. 20% normalized dataset is used for testing the prediction model.

2.7 Accuracy

Confusion matrix as depicted in Table 2 is used to evaluate the prediction model. Accuracy of the model is calculated using the following formula.

$$\text{Accuracy} = \frac{TN + TP}{TN + TP + FP + FN} \tag{5}$$

Table 2. Confusion matrix

Actual values	Predicted values	
	Positive (1)	Negative (0)
Positive (1)	True positive (TP)	False negative (FN)
Negative (0)	False positive (FP)	True negative (TN)

2.8 Procedure of the Proposed Work

Preprocess

1. Missing values in each feature of the dataset except number of times pregnant are replaced with the mean value of the corresponding feature of the dataset having the same diagnosed class.
2. Perform normalization on each feature value x using x/x_{max} where x_{max} is the maximum value of the corresponding feature.

Feature Selection

1. Construct single hidden layer neural network.
2. Train the neural network using standard backpropagation.
3. Prune the Input layer using N2PS algorithm.

Prediction

1. Sort the value under every pruned feature of the dataset.
2. Divide each feature of the dataset into five different groups based on its risk factors.
3. Find probability for each data under every feature using (3).
4. Find weight value for each data using (4).
5. Apply KNN on the resultant weighted values for prediction.

3 Experimental Results

The missing value of glucose, blood pressure, skin thickness, insulin and BMI are replaced with the mean value of the corresponding feature of the dataset having the same diagnosed class and is shown in Table 3. The values under all eight features are normalized.

The constructed architecture of the neural network is 9-13-1 which includes 1 bias neuron in the input layer. The number of hidden neuron is selected based on trial and error. Sigmoidal activation function is used in the hidden and output layers. The weights between the input layer and hidden layer and the weights between the hidden layer and output layer are initialized with the random numbers from the range $[-1\ 1]$. The network is trained upto 10,000 epochs and λ value is set as 0.09 by trial and error. After 25 different trials, network training error is 0.037706 mean squared error in average.

Next, the N2PS pruning algorithm is applied on the trained network for selecting the significance features. Table 4 shows the obtained s_i value of the features during pruning.

Table 3. Replacement values of the missing values

Feature name	Non-diabetic data		Diabetic data	
	No. of missing	Replacement value	No. of missing	Replacement value
Glucose	3	110	2	141
Blood pressure	19	68	16	71
Skin thickness	139	20	88	22
Insulin	236	69	138	100
BMI	9	30	2	35
Diabetes pedigree function	0	–	0	–
Age	0	–	0	–

Table 4. s_i value of features

Feature	s_i value
No. of times pregnant	2262.393715079818
Glucose	6130.979441303143
Blood pressure	5934.897467085981
Skin thickness	2637.893695983289
Insulin	1328.917737959378
Body mass index	4821.534577048848
Diabetes pedigree function	1955.4308894864764
Age	4121.510150165891

The calculated threshold alpha is 3244.505681524072. Since the s_i value of glucose, blood pressure, BMI and age are greater than alpha that features are selected as significant features and are only considered for further processing. Table 5 shows the selected range in each feature for calculating probability of data inside the range.

Table 5. Data range of features

Glucose	
Feature value	No. of records
<=90	110
>90 && <=100	99
>100 && <=120	208
>120 && <=140	157
>140	194

Blood pressure	
Feature value	No. of records
<=60	123
>60 && <=70	236
>70 && <80	204
>=80 && <90	145
>=90	60

BMI	
Feature value	No.of records
<=25	112
>25 && <=30	189
>30 && <=35	223
>35 && <=40	148
>40	96

Age	
Feature value	No. of Records
<=25	267
>25 && <=30	150
>30 && <=40	157
>40 && <=50	113
>50	81

The probability for each data in the group is calculated by considering size of the neighbor as 7. Here neighbor(k) means they occur nearer by their index in the sorted data of the feature and not by their distance. The different k values are considered and the results obtained are tabulated in Table 6. Number 7 is decided as the optimal k value. Sensitivity tells the correct disease status of the patient. An example for probability calculation is given in the Table 7. The data given in the Table 5 are taken from each group under the glucose feature.

Table 6. Obtained results for different k values

k	Sensitivity	Specificity
5	94.117645	5.882353
6	87.5817	12.418301
7	98.039215	1.9607844
8	96.732025	3.2679741

Table 7. Sample weight calculation

Feature Name	S.No	Feature value x_i	Target class value	Neighbors	# of data having same diagnosis category of x_i	Probability value	Weighted feature value
Glucose	1	0.417085	0	2,3,4,5,6,7,8	6	0.857143	0.357501
	2	0.422111	0	1,3,4,5,6,7,8	6	0.857143	0.361809
	3	0.497487	0	1,2,4,5,6,7,8	6	0.857143	0.426417
	4	0.502513	0	1,2,3,5,6,7,8	6	0.857143	0.430725
	5	0.592965	1	2,3,4,6,7,8,9	1	0.142857	0.084709
	6	0.59799	0	3,4,5,7,8,9,10	4	0.571429	0.341709
	7	0.688442	0	3,4,5,6,8,9,10	4	0.571429	0.393395
	8	0.693467	0	3,4,5,6,7,9,10	4	0.571429	0.396267
	9	0.959799	1	3,4,5,6,7,8,10	2	0.285714	0.274228
	10	0.969849	1	3,4,5,6,7,8,9	2	0.285714	0.2771

The 80% of the weighted dataset is given to KNN prediction model and the number of neighbor (k) for the KNN is considered as 15 by trial and error. Table 8 shows different k values and its obtained results. The 20% non-weighted normalized data is used for testing and it gives classification accuracy as 95.92%.

Table 8. Obtained sensitivity for different k values

k	Sensitivity
7	94.117645
8	90.84967
9	95.424835
10	94.77124
11	96.07843
12	95.424835
13	95.424835
14	97.38562
15	98.039215
16	96.07843
17	97.38562
18	94.77124
19	96.07843
20	92.813455

The confusion matrix obtained from the best cross validation experiment is given in Table 9. Table 10 shows the comparison of the predicted accuracy obtained by various researchers for the PID dataset. When the same experiment is carried without pruning the features, the PID dataset gives only 90.19% prediction accuracy which is lesser than the accuracy of the proposed method.

Table 9. Obtained confusion matrix predicted values

	Diabetic	Non-diabetic
Diabetic	51	3
Non-diabetic	0	99

Table 10. Comparison of PID dataset accuracy

Authors	Accuracy achieved
Alehegan et al. [4]	90.36% using proposed ensemble method
Yahyaoui et al. [7]	76.81% using convolutional neural network 65.38% using SVM 83.67% using random forest
Mujumdar et al. [11]	76% using logistic regression 77% using gradient boost classifier 77% using LDA 77% using adaboost classifier 76% using extra trees classifier 67% using gaussian NB 75% using bagging 72% using random forest 74% using Decision tree 67% using Perceptron 68% using SVC 72% using KNN
Tigga et al. [1]	74.4% using logistic regression 70.8% using KNN 74.4% using SVM 68.9% using Naïve Bays 69.7% using decision tree 75% using random forest
Gupta et al. [15]	85.06% using KNN
Hasan et al. [5]	95% using proposed ensemble model of AdaBoost + XGBoost
The proposed model	95.92% using KNN

4 Conclusion

Proposed method constructs a model for predicting type 2 diabetic patients. Data are normalized after missing values in the dataset are replaced with the mean value of the corresponding feature of the diagnosed diabetes type patients. Significant features are pruned using N2PS algorithm. The data of the extracted significant features are replaced with its calculated weight value and KNN is used for predicting the diabetes. The comparative results of the experimental result show that the proposed method produces better accuracy for predicting type 2 diabetes.

References

1. Tigga, N.P., Garg, S.: Prediction of type 2 diabetes using machine learning classification methods. Procedia Comput. Sci. **167**, 706–716 (2020). https://doi.org/10.1016/j.procs.2020.03.336

2. Zhou, H., Myrzashova, R., Zheng, R.: Diabetes prediction model based on an enhanced deep neural network. EURASIP J. Wirel. Commun. Netw. **2020**(1), 1–13 (2020). https://doi.org/10.1186/s13638-020-01765-7

3. Shiva, Reddy, S., Sethi, N., Rajender, R.: A comprehensive analysis of machine learning techniques for incessant prediction of diabetes mellitus. Int. J. Grid Distrib. Comput. **13**, 1–22 (2020). https://doi.org/10.33832/ijgdc.2020.13.1.01

4. Alehegn, M., Joshi, R., Mulay, P.: Analysis and prediction of diabetes mellitus using machine learning algorithm. Int. J. Pure Appl. Math. **118**, 871–878 (2018)

5. Hasan, M.K., Alam, M.A., Das, D., Hossain, E., Hasan, M.: Diabetes prediction using ensembling of different machine learning classifiers. IEEE Access **8**, 76516–76531 (2020). https://doi.org/10.1109/ACCESS.2020.2989857

6. Zou, Q., Qu, K., Luo, Y., Yin, D., Ju, Y., Tang, H.: Predicting diabetes mellitus with machine learning techniques. Front. Genet. **9**, 1–10 (2018). https://doi.org/10.3389/fgene.2018.00515

7. Anwar, F., Qurat-Ul-Ain, Ejaz, M.Y., Mosavi, A.: A comparative analysis on diagnosis of diabetes mellitus using different approaches – a survey. Inform. Med. Unlocked **21**, 100482 (2020). https://doi.org/10.1016/j.imu.2020.100482

8. Bukhari, M.M., Alkhamees, B.F., Hussain, S., Gumaei, A., Assiri, A., Ullah, S.S.: An improved artificial neural network model for effective diabetes prediction. Complexity **2021**, 1–10 (2021). https://doi.org/10.1155/2021/5525271

9. Sneha, N., Gangil, T.: Analysis of diabetes mellitus for early prediction using optimal features selection. J. Big Data **6**(1), 1–19 (2019). https://doi.org/10.1186/s40537-019-0175-6

10. Kopitar, L., Kocbek, P., Cilar, L., Sheikh, A., Stiglic, G.: Early detection of type 2 diabetes mellitus using machine learning-based prediction models. Sci. Rep. **10**, 1–12 (2020). https://doi.org/10.1038/s41598-020-68771-z

11. Muhammad, L.J., Algehyne, E.A., Usman, S.S.: Predictive supervised machine learning models for diabetes mellitus. SN Comput. Sci. **1**, 1–10 (2020). https://doi.org/10.1007/s42979-020-00250-8

12. Maniruzzaman, M., Rahman, M.J., Ahammed, B., Abedin, M.M.: Classification and prediction of diabetes disease using machine learning paradigm. Health Inf. Sci. Syst. **8**(1), 1–14 (2020). https://doi.org/10.1007/s13755-019-0095-z

13. Gupta, S.C., Goel, N.: Enhancement of performance of k-nearest neighbors classifiers for the prediction of diabetes using feature selection method. In: 2020 IEEE 5th International Conference Computer Communication Automation ICCCA 2020, pp. 681–686 (2020). https://doi.org/10.1109/ICCCA49541.2020.9250887

14. Le, T.M., Vo, T.M., Pham, T.N., Dao, S.V.T.: A novel wrapper-based feature selection for early diabetes prediction enhanced with a metaheuristic. IEEE Access **9**, 7869–7884 (2021). https://doi.org/10.1109/ACCESS.2020.3047942

15. Ahmad, H.F., Mukhtar, H., Alaqail, H., Seliaman, M., Alhumam, A.: Investigating health-related features and their impact on the prediction of diabetes using machine learning. Appl. Sci. **11**, 1–18 (2021). https://doi.org/10.3390/app11031173

16. Gethsiyal Augasta, M., Kathirvalavakumar, T.: A novel pruning algorithm for optimizing feedforward neural network of classification problems. Neural Processing Lett. **34**(3), 241–258 (2011). https://doi.org/10.1007/s11063-011-9196-7

17. Gao, Y., Gao, F.: Edited AdaBoost by weighted kNN. Neurocomputing **73**, 3079–3088 (2010). https://doi.org/10.1016/j.neucom.2010.06.024

18. Gou, J.: A New Distance-weighted k-nearest Neighbor Classifier A New Distance-weighted k-nearest Neighbor Classifier (2016)

Harnessing Energy of M-ary Hopfield Neural Network for Connectionist Temporal Sequence Decoding

Vandana M. Ladwani[1,2]([⊠]) [ID] and V. Ramasubramanian[1]([⊠]) [ID]

[1] International Institute of Information Technology - Bangalore, Bangalore, India
{vandana.ladwani,v.ramasubramanian}@iiitb.ac.in
[2] PES University, Bangalore, India

Abstract. Sequence decoding is the core component of systems that deal with sequence alignment problems like continuous speech recognition, visual scene labelling, multimedia storage and retrieval. In this paper, we address the problem of connectionist temporal sequence decoding in the context of movie sequence data. We present a novel decoding algorithm and associated system which harnesses the energy of an extended M-ary Hopfield Neural Network (MHNN) to decode movie sequence data. This is a first of its kind approach using the Hopfield associative memory based connectionist framework to solve the sequence alignment problem. We demonstrate with experiments the robustness of the proposed algorithm and system to handle temporal variability of the input sequence which is very crucial for connectionist sequence decoding system.

Keywords: Extended M-ary Hopfield Neural Network · Connectionist temporal sequence decoding · Auto-associative memory

1 Introduction

Connectionist temporal sequence decoding is the process of transcribing an input sequence of data (e.g. vector sequences or image sequences in a movie or multimedia content) into an optimal sequence of discrete labels drawn from a vocabulary of sub-sequence labels, and also deriving the segment boundaries in the input sequence associated with each label. This thus essentially involves an automatic and optimal 'segmentation and labeling' task of the input sequence, in such a way to match the original ground truth label sequence and the corresponding segment boundaries of each label as closely as possible. This problem is central in real world applications such as continuous speech recognition, movie scene recognition, gesture recognition etc. Sequence decoding involves two sub-tasks

1. Segmenting unsegmented data which may suffer from temporal variability due to insertion,deletion and substitution errors
2. Label segmented data

© Springer Nature Switzerland AG 2022
R. Chbeir et al. (Eds.): MIKE 2021, LNAI 13119, pp. 13–23, 2022.
https://doi.org/10.1007/978-3-031-21517-9_2

Hopfield network is an auto-associative memory formulation originally proposed as a content addressable retrieval system for bipolar data [1]. Hopfield network with graded neurons is well suited for real world data [2]. The original Hopfield network was trained using Hebbian learning which permits $O(0.14N)$ storage capacity for uncorrelated data [3]. Hopfield network trained with Psuedo-inverse learning rule allows $O(N)$ storage capacity [4]. Shriwas et al. proposed Hopfield based Auto-associative framework for multi-modal data storage and retrieval which involves static data [5]. Bijjani proposed M-ary Hopfield Neural Network for content based retrieval of static gray images [6]. Our work focuses on building a M-ary Hopfield Neural Network (MHNN) based decoder system which uses the energy of the network to realize connectionist temporal sequence decoding. We experimentally demonstrate the performance of the proposed system for unsegmented and unlabeled continuous sequences composed of movie clips which are composed of temporally different sub-sequences of the stored prototypes; results for two randomly generated test sequences are presented in the paper.

2   Hopfield Network Based Formulations

2.1   Basic Hopfield Network for Static Patterns

Hopfield network is a fully connected network consisting of N neurons. Hopfield network possesses the interesting property of 'auto-association' - where, given a state which can be clean/corrupted/partial version of a 'stored pattern', the network can retrieve the complete 'stored pattern' [7]. The original Hopfield network is based on McCulloch-Pitts model, where each neuron can assume two states on/off (+1 and −1 for bipolar case) and the state of the network at a particular instance is represented as $\mathbf{x} = [x_1, x_2, ..x_n.., x_N]^T$. Functioning of the network can be described in following two phases.

Storage Phase. In this phase, network memorizes the patterns ('Prototypes') presented to it using learning rules. Hebbian learning and Pseudo Inverse learning are two such examples [3,4]. Let $\xi^1, \xi^2, \xi^3, \ldots, \xi^u, \ldots, \xi^U$ represent U patterns each of length N to be stored in the network. The coupling strength (or synaptic weight) from i^{th} to j^{th} neuron is given as

1. Hebbian learning rule:

$$W_{ij} = \frac{1}{N} \sum_{u=1}^{U} \xi_i^u \xi_j^u \qquad (1)$$

2. Pseudo Inverse learning rule:

$$W_{ij} = \frac{1}{N} \sum_{k,l} \xi_i^k C^{-1} \xi_j^l \qquad (2)$$

where, C represents the co-variance matrix of the training vectors.

Retrieval Phase. In this phase, the Hopfield network trained as above is presented with probe vector ξ^{probe}. Each neuron updates the state as per the given rule until convergence.

$$h_i(t) = \sum_{j=1}^{N} W_{ij} x_j(t); \qquad x_i(t+1) = sign(h_i(t)); \qquad x_j(0) = \xi_j^{probe} \qquad (3)$$

where, $h_i(t)$ represents input potential to the ith neuron. The state update can be performed synchronously where all neurons update their states at the same time or asynchronously where neurons update their states in a random, but sequential, manner. Each state can be characterised by a metric termed as 'energy', which is unique to a particular state. The energy associated with a state is given by

$$E(\mathbf{x}) = -\frac{1}{2} \sum_{i=1}^{N} \sum_{j=1}^{N} W_{ij} x_i x_j \qquad (4)$$

The retrieval process can be visualised as a descent in the 'energy landscape' from the start state (noisy version of an input stored pattern or prototype pattern) within a basin of attraction to the bottom of the respective well (basin) which represents the 'prototype pattern' memorised.

2.2 M-ary Hopfield Network for Static Patterns

In M-ary Hopfield Network (MHNN), each neuron can assume any of the permissible M values at a particular instance. Bijjani and Das proposed M-ary Hopfield Network trained using Hebbian learning rule for storing gray-scale images [6]. They demonstrated retrieval of stored images from MHNN with corrupted query images under asynchronous update rule with staircase activation function.

2.3 Hopfield Based Network for Sequence Storage and Retrieval

In order to store and retrieve isolated temporal sequences, the network needs to memorize static information which enables the network to stabilize while in a particular state and transition information which helps the network to switch from one state to the next of the stored sequence. Zhang et al. adapted a continuous Hopfield network with tanh activation function to store and retrieve sequence data [8]. Let matrix Ξ with column vectors $\xi^1, \xi^2, \xi^3, \ldots, \xi^l \ldots, \xi^L$ represent a sequence of length L to be stored; here, ξ^l represents the lth state in the sequence. They proposed Pseudo-inverse based learning rule as follows to enable the Hopfield network to store sequence data

$$J = \frac{\beta_k}{2}(C_0 \Xi \Xi^+ + C_1 \Xi_1 \Xi^+) \qquad (5)$$

$\Xi_1 = (\xi^2, \xi^3, \ldots, \xi^l \ldots, \xi^L, \xi^1)$, Ξ^+ represents the Pseudo inverse of Ξ, β_k, C_0 and C_1 are constants. They presented results for storing short synthetic sequences of states where each state vector consists of bipolar elements.

3 M-ary HNN for Sequence Storage and Retrieval

We adapt M-ary Hopfield network proposed by Bijjani as described in Sect. 2.2 to label isolated sequences [6]. We use Pseudo-inverse based dual weight learning rule to store a sequence of specific length L, $\xi^1, \xi^2, \xi^3, \ldots, \xi^l, \ldots, \xi^L$ in a particular M-ary HNN. Dual weight learning is a two stage one-shot learning method to memorize individual states and state transition information. Dual weight learning stores information about fixed point states and state transitions in separate weight matrices in contrast to single weight matrix in Zhang's work [8].

3.1 Dual Weight Learning

Static and Transition matrices are determined by Dual weight learning using the following rules,

$$S_{ij} = \frac{1}{N} \sum_{k,l} \xi_i^k C^{-1} \xi_j^l \tag{6}$$

$$T_{ij} = \frac{1}{N} \sum_{no\ of\ cycles} \Xi_1 \Xi^+ \tag{7}$$

where S_{ij} and T_{ij} respectively denote static and transition coupling strength between i^{th} and j^{th} unit, Ξ_1 is rotated version of Ξ .

3.2 Isolated Sequence Retrieval from M-ary HNN

Given a query state (clean/corrupted) as a trigger state, an iterative procedure is used to retrieve the stored sequence containing the query state. Each step consists of a two stage procedure 'synchronous firing' to switch from one state to another and 'asynchronous firing' to stabilize the current state. The system dynamics are governed by the equations as follows.

$$h_i(t) = \sum_{j=1}^{N} S_{ij} x_j(t); \qquad x_i(t+1) = f(h_i(t)) \tag{8}$$

$$h_i(t) = \sum_{j=1}^{N} T_{ij} x_j(t); \qquad x_i(t+1) = f(h_i(t)) \tag{9}$$

where, $h_i(t)$ represents input potential to the ith neuron, $x_i(t+1)$ is the updated state of ith neuron and $f()$ is a staircase activation function.

Analogous to Ising model, the energy function is used to encode the state of the system. Thus if MHNN is triggered with corrupted/clean version of any particular state (start/intermediate) of a sequence it has memorized and is set to run for the duration equal to the length of the memorized sequence, the network transits form one well (basin of attraction) to another in the energy landscape as shown in Fig. 1 to retrieve the complete sequence. The energy landscape is unique for a particular sequence. The energy associated with the sequence is determined as per the equation

$$E_{seq} = \frac{1}{L} \sum_{i=1}^{L} E_{state_i} \tag{10}$$

where,
E_{seq} represents energy of sequence of length L
E_{state_i} is the energy of ith state given by the equation

$$E_{state_i} = -\frac{1}{2} \sum_{l=1}^{N} \sum_{m=1}^{N} S_{lm} x_l x_m \tag{11}$$

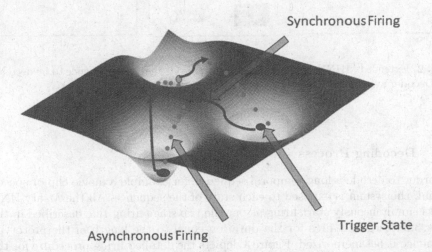

Fig. 1. Retrieval dynamics: 1) Asynchronous firing: network transits to bottom of the well (stable state in the basin of attraction), 2) Synchronous firing: system transits to shallow region which marks beginning of next state in the sequence

4   Connected Temporal Sequence Decoding Using Extended M-ary HNN

4.1   Model

The 'extended M-ary HNN' based decoder consists of multiple M-ary HNNs, each MHNN unit trained to store prototype sequence as described in Sect. 3. Figure 2(b) shows decoder consisting of K MHNN units and the respective memorised clips. As shown in Fig. 2(a), the input to the system is a connected sequence which consists of 'temporally variable' versions of the stored prototypes which may result due to insertion/deletion and substitution or vector errors. Each MHNN unit is sensitive to the region of the given sequence which resembles within some tolerance to the prototype it has memorised. The system outputs segment boundaries and associated labels as shown in Fig. 2(c).

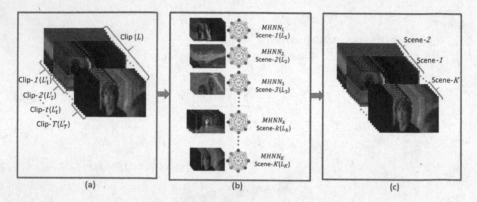

Fig. 2. Extended MHNN based decoder system: (a) Connected sequence to be decoded (b)Decoder system with K MHNN units (C) Segmented clips with labels

4.2   Decoding Process

In order to decode a long temporal sequence, for example a movie clip or speech signal, the system is exposed to each state of the sequence. All the M-ary HNN units simultaneously start firing as per the two stage firing rule described in the Sect. 3.2. Each unit fires for the duration equal to the length of the prototype sequence it has memorized. Figure 3 depicts the detailed firing procedure for the unit MHNN1 of the decoder system in Fig. 2. Proposed decoder system keeps track of the average energy of retrieved sequence shown in Fig. 3(c), for each M-ary HNN during the firing process. Figure 3(d) shows the color coded average energy column vector for MHNN1, where each element (blue color indicates low and red color high values) corresponds to energy of the retrieved sequence at a particular timestamp. Figure 3(e) shows the average energy profile of MHNN1, where red circled value indicates the energy of retrieved sequence when triggered with state at the 2nd timestamp in the input sequence (high energy) and blue circled value indicates the energy of the retrieved sequence at the 7th timestamp (low energy).

The retrieved energy profile for a particular M-ary HNN can be represented by the following matrix

$$\begin{pmatrix} E_{11}^k & E_{12}^k & E_{13}^k & \cdots & E_{1j}^k & \cdots & E_{1L_k}^k & E_1^k \\ E_{21}^k & E_{22}^k & E_{23}^k & \cdots & E_{2j}^k & \cdots & E_{2L_k}^k & E_2^k \\ E_{31}^k & E_{32}^k & E_{33}^k & \cdots & E_{3j}^k & \cdots & E_{3L_k}^k & E_3^k \\ \vdots & \vdots & \vdots & \ddots & \vdots & \ddots & \vdots & \vdots \\ E_{i1}^k & E_{i2}^k & E_{i3}^k & \cdots & E_{ij}^k & \cdots & E_{iL_k}^k & E_i^k \\ \vdots & \vdots & \vdots & \ddots & \vdots & \ddots & \vdots & \vdots \\ E_{L1}^k & E_{L2}^k & E_{L3}^k & \cdots & E_{Lj}^k & \cdots & E_{LL_k}^k & E_L^k \end{pmatrix},$$

where, E_{ij}^k represents energy at i^{th} timestamp of input sequence for j^{th} state of the kth MHNN and E_i^k (last column in the matrix) represents average energy of sequence retrieved from k^{th} MHNN when triggered with i^{th} state of the sequence to be decoded.

As observed in Fig. 3(e), the average energy profile has a region with low energies, which corresponds to the sequence clip to which MHNN unit under consideration (MHNN1) has responded; peak picking procedure on first order derivative of average energy profile, as shown in Fig. 3(f), is used to segment the identified region.

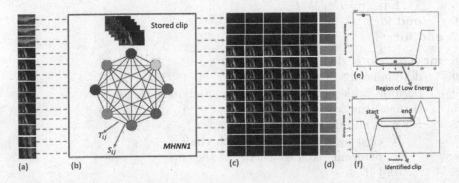

Fig. 3. Decoding Procedure: (a) Unsegmented movie sequence to be decoded, (b) MHNN1 in action, (c) Retrieved sequence at each timestamp, (d) Heat map for average energy of MHNN1 at each timestamp, (e) Average energy profile of MHNN1, (f) Segmented clip (Color figure online)

In order to segment a given temporal sequence into distinct clips with clearly identified start and end boundaries, we propose M-ary HNN energy based decoding algorithm (Algorithm 1).

5   Dataset and Experiments

We synthesized temporal sequences from movie clips extracted from the movie 'Lord of the Rings'. Clips of duration 1 s, 1.6 s and 2 s are taken from start,

---

Algorithm 1. Connected Sequence Decoding using Extended M-ary HNN

Input: $\xi^1, \xi^2, \xi^3, \ldots, \xi^l, \ldots, \xi^L$ (temporal sequence of length L to be decoded)

Output: $(b_1, e_1, label_1)$, $(b_2, e_2, label_2), \ldots, (b_t, e_t, label_t), \ldots, (b_T, e_T, label_T)$(labelled segments)

1: **for** $i \leftarrow 1$ to L **do**   ▷ each state in temporal sequence
2:     **for** $k \leftarrow 1$ to K **do**   ▷ each MHNN unit in the system
3:         $(x_1, x_2, \ldots, x_l, \ldots, x_m, \ldots, x_N) \leftarrow (\xi_1^i, \xi_2^i, \ldots, \xi_l^i, \ldots, \xi_m^i, \ldots, \xi_N^i)$ ▷
   Initialize MHNN with current state
4:         **for** $j \leftarrow 1$ to L_k **do** ▷ Fire network using two stage firing procedure (section 3.2)
5:             $(E_{ij}^k \leftarrow -\frac{1}{2} \sum_{l=1}^{N} \sum_{m=1}^{N} S_{lm} x_l x_m)$ ▷ Energy of retrieved state
6:         **end for**
7:     **end for**
8: **end for**

9: **for** $k \leftarrow 1$ to K **do**   ▷ for each MHNN unit
10:     **for** $i \leftarrow 1$ to L **do**
11:         $Eavg_i^k \leftarrow 0$
12:         **for** $j \leftarrow 1$ to L_k **do**
13:             $Eavg_i^k = Eavg_i^k + E_{ij}^k$
14:         **end for**
15:         $Eavg_i^k = Eavg_i^k / L_k$
16:     **end for**
17: **end for**

18: **for** $k \leftarrow 1$ to K **do**
19:     determine the derivative of $Eavg_i^k$
20:     perform peak picking on the differentiation sequence
21:     segment the sequence into clips
22:     **for** each clip **do**
23:         **if** cliplength > thresholdlength **then**
24:             determine average energy of the clip
25:         **end if**
26:     **end for**
27:     determine the clip with the minimum energy $clip_{min}$
28:     **if** $E(clip_{min}) - E(MHNN_k) <$ thresholdenergy **then**
29:         mark start and end boundaries and label the clip with id of $MHNN_k$
30:     **end if**
31: **end for**

---

mid and end of the movie termed as Clip-1, Clip-2 and Clip-3 respectively. Frames are extracted at the rate of 25 frames/sec; RGB frames are converted to gray frames (8 bit encoding) with resolution set to 100×100 using nearest neighbor interpolation; each rasterized frame corresponds to a specific state in the sequence. We built a decoder system with three MHNNs, MHNN1, MHNN2 and MHNN3 each with network size, $N = 10000$ neurons. The parameter M is set to 256, i.e., each neuron can assume any value from a set of 256 values. MHNN1, MHNN2 and MHNN3 are trained using dual weight learning to store Clip-1, Clip-2 and Clip-3 respectively as 'prototype sequences'. We consider two scenarios here to demonstrate working of the proposed decoder system. Table 1 lists the prototype sequence information for each MHNN.

Table 1. Stored prototype sequence details

Decoder unit	Stored clip	Stored clip-id	Clip length	Clip duration
MHNN1	Clip-1	Scene-1	25 frames	1 s
MHNN2	Clip-2	Scene-2	40 frames	1.6 s
MHNN3	Clip-3	Scene-3	50 frames	2 s

5.1 Scenario-A

In this experiment, we concatenate Clip-3', Clip-2' and Clip-1' which are corrupted versions of Clip-3, Clip-2 and Clip-1 with 20% insertion/deletion and substitution errors to construct test sequence: Sequence-A. The proposed energy based decoding as illustrated in 'Algorithm 1' is performed with the constructed MHNN decoder. Energy profiles of 3 MHNNs which represent average energy of specific MHNN at a particular instance are shown in Fig. 4. Table 2 shows the results of the experiment.

Table 2. Decoding results for Sequence-A

Ground truth					Decoder output		
Input sequence	Start	End	Id	I/D/S	(start, end, id)	Identified by	Energy minima
Segment-1(54)	0	53	Scene-3	20%	(0, 53, Scene-3)	HNN3(50)	−3.82E+07
Segment-2(44)	54	97	Scene-2	20%	(54, 97, Scene-2)	HNN2(40)	−3.05E+07
Segment-3(28)	98	125	Scene-1	20%	(98, 125, Scene1)	HNN1(25)	−1.47E+07

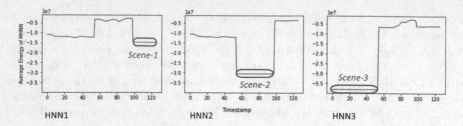

HNN1   HNN2   Timestamp   HNN3

Fig. 4. Decoding continuous movie sequence (Sequence-A)

5.2   Scenario-B

In Scenario-B, we performed decoding of a long temporal sequence: Sequence-B, constructed by concatenating Clip-2′, Clip-1′, Clip-3′, Clip-1″, Clip-2″ and Clip-3″ where Clip-1′, Clip-2′ and Clip-3′ are same as in Scenario-A. Clip-1″, Clip-2″ and Clip-3″ are synthesized from Clip-1, Clip-2 and Clip-3 by 30% insertion/deletion and substitution errors randomly injected in the prototype clips. Figure 5 shows the average energy profile for each MHNN. Results of this experiment are presented in the Table 3. It can be seen from the results of the experiments, that the system segments and labels given temporal sequence with clearly identified start and end boundaries with no induced error. The system efficiently deals with the 'alignment problem' which is an important aspect to be dealt with for continuous temporal sequence decoding.

Table 3. Decoding results for Sequence-B

Ground truth					Decoder output		
Input sequence	Start	End	Id	I/D/S	(start, end, id)	Identified by	Energy minima
Segment-1	0	45	Scene-2	20%	(0, 45, Scene-2)	HNN2(40)	−3.05E+07
Segment-2	46	72	Scene-1	20%	(46, 72, Scene-1)	HNN1(25)	−1.47E+07
Segment-3	73	127	Scene-3	20%	(73, 127, Scene-3)	HNN3(50)	−3.82E+07
Segment-4	128	154	Scene-1	30%	(128, 154, Scene-1)	HNN1(25)	−1.47E+07
Segment-5	155	196	Scene-2	30%	(155, 196, Scene-2)	HNN2(40)	−3.05E+07
Segment-6	197	251	Scene-3	30%	(197, 251, Scene-3)	HNN3(50)	−3.82E+07

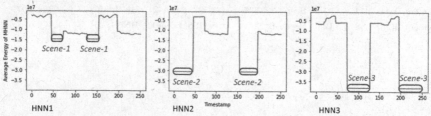

HNN1   HNN2   Timestamp   HNN3

Fig. 5. Decoding continuous movie sequence (Sequence-B)

6   Conclusions

We have proposed an extended M-ary HNN based decoder system. We observed that the M-ary HNN responds only to the regions of temporal sequences which match the prototype sequence it has memorised with low energy which is the core of the proposed decoding system. The system harnesses the energy of a set of M-ary HNNs to efficiently carry out decoding of a given temporal sequence. This has never been attempted so far. The proposed system is robust to insertion, deletion and substitution errors which is a very important requirement to handle temporal sequences which vary in length with respect to the prototype sequences along with pattern variability.

References

1. Hopfield, J.J.: Neural networks and physical systems with emergent collective computational abilities. Proc. Natl. Acad. Sci. **79**(8), 2554–2558 (1982)
2. Hopfield, J.J.: Neurons with graded response have collective computational properties like those of two-state neurons. Proc. Natl. Acad. Sci. **81**(10), 3088–3092 (1984)
3. Hebb, D.O.: The Organization of Behavior: A Neuropsychological Theory. Wiley, New York (1949). 335 p. Sci. Educ. **34**(5), 336–337 (1950)
4. Personnaz, L., Guyon, I., Dreyfus, G.: Collective computational properties of neural networks: new learning mechanisms. Phys. Rev. A **34**(5), 4217 (1986)
5. Shriwas, R., Joshi, P., Ladwani, V.M., Ramasubramanian, V.: Multi-modal associative storage and retrieval using Hopfield auto-associative memory network. In: Tetko, I.V., Kůrková, V., Karpov, P., Theis, F. (eds.) ICANN 2019. LNCS, vol. 11727, pp. 57–75. Springer, Cham (2019). https://doi.org/10.1007/978-3-030-30487-4_5
6. Bijjani, R., Das, P.: An M-ary neural network model. Neural Comput. **2**(4), 536–551 (1990)
7. Haykin, S.: Neural Networks: A Comprehensive Foundation, 2nd edn. Prentice Hall PTR, USA (1998)
8. Zhang, C., Dangelmayr, G., Oprea, I.: Storing cycles in Hopfield-type networks with pseudoinverse learning rule: admissibility and network topology. Neural Netw. **46**, 283–298 (2013)

Integrative Analysis of miRNA-mRNA Expression Data to Identify miRNA-Targets for Oral Cancer

Saswati Mahapatra[1]([⊠]) [iD], Rajendra Prasath[2] [iD], and Tripti Swarnkar[1] [iD]

[1] Department of Computer Application, Siksha O Anusandhan Deemed to be University, Bhubaneswar, India
{saswatimohapatra,triptiswarnkar}@soa.ac.in
[2] Department of Computer Science and Engineering, Indian Institute of Information Technology, Sri City, Chittoor, India
rajendra.prasath@iiits.in

Abstract. Micro RNAs (miRNAs) are small non coding RNA sequences consisting of 20–23 nucleotides that govern the post transcriptional expression of genes in both normal and disease condition of the cell. Thus, identification of most influencing miRNAs and the associated mRNAs becomes a research quest in diagnostic and prognostic application of cancer. In this study we conducted an integrated analysis of Next Generation Sequencing based miRNA-mRNA expression data to identify dysregulated miRNAs and their target mRNAs for Oral Cancer. A sensible combination of datamining tools such as Random Forest (RF), K-nearest Neighbour (KNN), Support Vector Machine (SVM), log-Fold Change, Adjusted p-values, Matthews coefficient correlation (MCC), Prediction accuracy was considered for this analysis. The prioritized cancer specific target genes obtained in this approach exhibited a MCC value of 0.9 and achieved a consistently higher prediction accuracy of 95% when subjected to classifiers RF, KNN and SVM. These target genes can be presented as predictive variables for early diagnosis of cancer. The selected miRNA-target genes can further be biologically validated to confirm their participation in disease specific pathways and biological processes.

Keywords: Micro RNA · miRNA-targets · miRNA-mRNA association · Next-generation sequencing · Oral Cancer

1 Introduction

Cancer has now ranked as leading cause of death worldwide [1]. Oral cancer (OC) is a category of head and neck squamous cell carcinoma, mostly developed on the lip, floor of the mouth, cheek lining, gingiva, palate or in the tongue. In India, OC is measured among top three types of cancers which accounts for more than 30% of all cancers [2]. Often OC is mostly diagnosed at its advanced stage i.e., when cancer has metastasized to another location, most likely the

© Springer Nature Switzerland AG 2022
R. Chbeir et al. (Eds.): MIKE 2021, LNAI 13119, pp. 24–32, 2022.
https://doi.org/10.1007/978-3-031-21517-9_3

lymph nodes of the neck, which results in low treatment outcomes and leaves patient with significantly low survival rate [3]. Alcohol addiction, practice of tobacco products like cigarettes, smokeless tobacco and viral infection are the most common risk factors for oral cancer [4].

Cancer is a multi-step process which causes due to mutation in genes that controls cell behaviour. Mutated genes may result in uncontrolled growth of cells that invade and cause the adjacent tissue impairment [5]. Micro RNAs (miRNAs) are small non coding RNA sequences consisting of 20–23 nucleotides that are incriminated in numerous biological, anatomical processes including cell differentiation, cell signalling, apoptosis, metastasis and response to infection [6]. Dysregulated expression pattern of miRNA is an indicator for initiation and progression of various disease including cancer [23]. Hence, identification of dysregulated miRNAs becomes crucial towards understanding of the biological mechanism behind miRNAs. miRNA govern the post transcriptional expression of genes by complementary base pairing with target m-RNAs in both normal and disease condition of the cell [7]. Thus, prediction of the miRNA-mRNA target interactions becomes significant to elucidate the mechanism by which miRNA act in carcinogenesis process. However, it has become a current challenge to correctly characterize the course of action of miRNAs on their mRNA targets, because each miRNA has multiple mRNA targets and vice versa [8]. This association of miRNA-mRNA targets highlights the importance of integrating miRNA expression with downstream mRNA target genes [9].

2   Background Study

Several studies have been carried out in literature to identify novel miRNA signatures associated with cancer and elucidate miRNA-mRNA target interactions. Seo et al. applied an integrative approach of miRNA, mRNA and protein expression data for identifying cancer-related miRNAs and investigating the gene-miRNA association [10]. Modules of highly correrated miRNA, mRNA and proteins were constructed using SAMBA bi-clustering algorithm and a Bayesian network model. The regulatory relationship between these modules were then investigated for precise analysis of miRNA-target gene interactions. Another integrative approach was proposed in [11] to identify the mRNA targets of abnormally regulated miRNAs. Several aberrantly expressed miRNAs and the associated target mRNA signatures were identified in this approach across six different cancer types. Sathipati et al. proposed SVM-HCC model based on inheritable bi-objective combinatorial genetic algorithm for selecting novel miRNA signatures for predicting hepatocellular carcinoma stages [12]. A hierarchical integrative model was utilized in [13] to uncover the miRNA-mRNA associations utilizing the sequence data of miRNA and mRNA. The identified miRNA-mRNA pairs were observed to be involved in processes contributing to hepatocellular carcinoma progression. A biphasic technique of machine learning based feature selection followed by survival analysis was applied in [14] to identify the most significant miRNA biomarkers for breast cancer subtype prediction. There is a strong association of miRNAs in various oral carcinomatous process. Thus, the abnormal expression detected in samples obtained from oral cancer patients are clinically significant in prediction and the development of effective treatments [16]. Falzole et al. utilized miRNA expression data set from GEO and TCGA miRNA profiling datasets to identify miRNAs signatures specific to OC [15].

In this study we proposed an integrated computational approach for identification and analysis of dysregulated miRNAs and their target mRNAs for Oral Cancer. Dysregulated miRNAs were prioritized based on their contribution in predicting the diseased condition. Further, putative dysregulated target mRNAs specific to cancer were identified and their prediction ability in separating the clinical conditions was examined.

The paper is organized as follows: Sect. 3 briefly describes the steps of our proposed method. Section 4 presents and discuss the empirical results of this study. Finally, Sect. 5 presents the conclusions of our study.

3   Materials and Methods

3.1   Dataset Used

Next-generation Sequencing based miRNA and mRNA expression data for the same patient were utilized in this work. The dataset was taken from GDC data portal of TCGA (https://portal.gdc.cancer.gov/). The Cancer Genome Atlas (TCGA) is a consortium of cancer genomics spanning over 33 cancer types which applies high throughput genome analysis techniques for characterizing genetic mutations responsible for cancer [17]. The data set consists of expression values of 1881 miRNAs and 18283 genes for 120 tumor samples and 44 matched normal samples.

3.2   Proposed Model

Figure 1 illustrates the workflow of our proposed model. The steps of our proposed work goes as follows:

(A) *Data preparation:* The data preparation step of our approach entailed removal of candidate miRNAs and genes with more than 30% of missing values followed by replacing the remaining missing values with mean of the sample [18]. Further, a logarithmic transformation base 2 [11] was applied on the resultant data in order to achieve normal distribution.

(B) *Identification of significant dysregulated miRNAs:* A differential expression analysis of miRNAs and genes was done to find significant miRNAs and genes that show quantitative changes in expression levels between experimental groups normal and diseased. The candidate miRNAs and genes were investigated with the help of adjusted p-values and log-transformed fold change for identifying dysregulated miRNAs. A change in expression profile was considered as filtering criteria for identifying differentially expressed miRNAs. miRNAs with adjusted p-value ≤ 0.05 and logFoldChange ≥ 2 [18] were considered to be significant in this study.

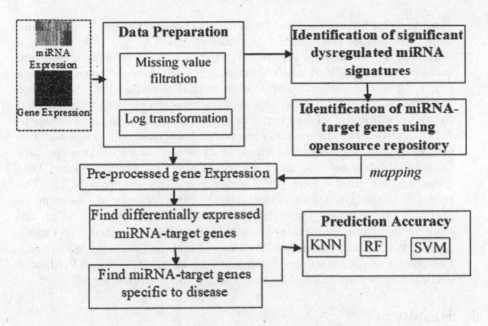

Fig. 1. Proposed model for identification of dysregulated miRNAs and associated target genes.

To evaluate the predictive ability of the differentially expressed miRNAs and to extract a handpicked of miRNAs, Random Forest (RF) classifier was adopted [11]. The RF is a learning method for classification, which works on the principle of ensemble learning by combining the solutions produced by multiple classifiers. The forest generated in the RF model consists of many decision trees of varying depths. RF method applies boot strap aggregating technique to train the decision trees. The samples left out during the training of each decision tree is referred as Out-Of-Bag (OOB) samples. For a new unseen sample, the learned RF model predicts by taking the average of the prediction outputs given by distinct decision trees. RF classifier can also be used to rank the features. Here we used Mean Decrease Accuracy (MDA) as the parameter for filtering significant miRNAs [24]. MDA is the proportion of observations that are incorrectly classified by removing the feature from the learned model [21]. The higher the MDA value, the more important the feature is.

(C) *Identification of miRNA-target genes using open source repository:* The potential targets for the dysregulated miRNAs resulted in the previous step, were obtained using the TargetScan database [19]. For finding the target genes, we considered the conserved miRNAs only. Because more than 60% of human genes contain targets of conserved miRNAs across species [20]. Top predicted target genes with the highest aggregate probability of conserved targeting (Aggregate P_{CT}), irrespective of site conservation were considered for further analysis. Finally, a set of distinct miRNA-target genes were identified for the inputted dysregulated miRNAs.

(D) *Screening of statistically significant target genes:* Differential expression analysis of the identified miRNA-target genes was done to screen the genes which express at different levels between clinical conditions. These genes are expected to offer precise biological insight into the processes affected by the condition(s) of interest. Furthermore, to unwrap the correspondence between the differentially expressed miRNA-target genes and disease of interest, a disease relatedness analysis was done by taking the data from NCBI (www.ncbi.nlm.nih.gov/geo/). Data of 8933 cancer-related (CR) genes and 316 oral cancer (OC) genes were obtained from NCBI. The screened target genes were investigated for the presence of cancer genes and oral cancer genes. Three different classifiers KNN (k = 3), SVM, RF were applied to examine the prognostic ability of the identified target genes against two clinical conditions. Parameters such as Specificity, Sensitivity, Precision, F-Score, Matthews coefficient correlation (MCC) and Prediction accuracy were used to measure the classification performance [22].

4   Results

We performed step wise analysis and selection of miRNA signatures and the associated target genes. The results obtained in this proposed work are presented on

(i) Prioritizing miRNA signatures
(ii) Identification of significant target genes specific to the disease
(iii) Effectiveness of the final selected target genes.

Table 1. Top 10 differentially expressed and computationally significant miRNA signatures.

miR_ID	MDA	LogFoldChange	AdjPvals
hsa-mir-130b	2.72	1.36	2.16454E−25
hsa-mir-99a	2.58	0.73	3.95543E−30
hsa-mir-101-2	2.28	0.85	3.85102E−29
hsa-mir-196b	2.21	1.82	1.5221E−22
hsa-mir-455	2.18	1.34	4.3197E−24
hsa-mir-101-1	2.10	0.85	8.32347E−29
hsa-mir-1301	2.00	1.94	4.8552E−20
hsa-mir-301a	1.95	2.13	1.48824E−16
hsa-mir-671	1.94	1.44	8.43309E−18
hsa-mir-503	1.75	2.27	0.000179404

(i) *Identification of prioritized miRNA signatures*

The miRNA expression and gene expression data collected from TCGA produced were inputted to the data preparation step of our proposed model which resulted in 493 miRNA signatures and expression values of 16478 genes. The differential expression analysis of the resultant 493 miRNA signatures resulted in 244 differentially expressed miRNA signatures with adjusted p-value <0.05 and logFoldChange >2. To further identify the miRNA signatures which are significant in disease prognosis, RF classifier was used. miRNA signatures were ranked in decreasing order of MDA value. miRNA signatures with MDA value >1 were considered significant. This resulted in 72 significant dysregulated miRNA signatures out of 244. This ensures that these handpicked 72 miRNA signatures are differentially expressed and computationally proficient as well. Among all, the top 10 significant miRNA signatures are illustrated in Table 1.

(ii) *Identification of significant target genes*

To obtain the potential targets for 72 notable miRNA signatures obtained in the previous step, a web-based target prediction tool: Target Scan was utilized. We systematically searched TargetScan for the identification of biological targets of the handpicked miRNAs. Target genes were queried for conserved miRNAs only. We obtained the top 50 targets for each conserved miRNAs with the highest aggregate probability of conserved targeting (PCT) value. It resulted in 1511 unique miRNA-target genes, which were finally mapped to pre-processed gene expression data obtained for OC. However, during mapping, a few miRNA-targets were dropped out because of its unavailability in the acquired TCGA gene expression data. This resulted in expression values of 1334 miRNA-target genes.

For selecting the target genes showing significant changes in different diseased conditions, the expression vector of 1334 target genes for tumor and normal samples were compared with respect to adjusted p-value and logarithmic fold change in expression levels. An adjusted p-value <0.05 and logFoldChange >2 was kept as cut off parameter. It resulted in 671 differentially expressed miRNA-target genes. The volcano plot in Fig. 2a clearly represents the differentially expressed miRNA-target genes marked with cyan colored dots.

Further, these target genes were reviewed for the existance of cancer-related genes and oral cancer genes using data collected from NCBI. Among 671 identified differentially expressed miRNA-targets, 331 genes were observed to be related to cancer whereas 19 genes were oral cancer genes. The proportion of CR and OC genes in the whole set of differentially expressed genes is demonstrated in Fig. 2b. These 350 (331 CR and 19 OC) genes were further validated in the following step with three different classifiers: KNN, RF and SVM.

(iii) *Effectiveness of the final selected target genes*

To examine the predictive efficiency of the 350 target genes obtained in the previous step, we run the classifiers KNN (k = 3), RF and SVM. All the classifiers were run with 10-fold cross validation. Table 2 illustrates the results

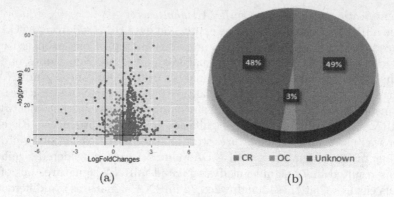

(a)   (b)

Fig. 2. a) Volcano plot showing differentially expressed miRNA-target genes with cyan coloured dots.(b) Presence of cancer-related (CR) and oral cancer (OC) genes in the selected group of miRNA-target genes.

Table 2. The classifier performance result of identified miRNA-target genes.

Classifiers	KNN (k = 3)		RF		SVM	
Parameters	T	N	T	N	T	N
Sensitivity	0.98	0.94	0.99	0.92	1.00	0.949
Specificity	0.95	0.98	0.92	0.99	0.95	1.00
Precision	0.98	0.94	0.97	0.97	0.98	1.00
MCC	0.98	0.94	0.92	0.92	0.96	0.96
F-Score	0.92	0.92	0.98	0.94	0.99	0.97
Accuracy %	95.6		97.1		98.5	

of classification. For all the three classifiers: KNN (k = 3), RF and SVM, the identified miRNA-target genes achieved an accuracy of 95.6%, 97.1% and 98.5% respectively. The acknowledged target features were observed with an average MCC value of 0.9 for the considered classifiers. A MCC value >0.5 is mostly considered to be significant in various machine learning platforms when there is an imbalanced ratio of input samples. The result shows that the distinguished target genes obtained in this approach can put a new light on OC prognosis with efficacy. These genes can further be biologically validated to confirm their participation in disease specific pathways and biological processes.

5   Conclusion

Identification of specific miRNAs and the associated target genes is crucial in characterizing the course of action of miRNAs in biological processes which lead towards cancer progression. The proposed work started with sample matched data of miRNA expression and gene expression to identify the dysregulated miRNAs and respective target genes. Up and down regulated miRNAs with high value for mean decrease in accuracy of the classifier were considered to be dysregulated. The specific top ranked target genes for the obtained dysregulated miRNAs were obtained from the online repository and further analyzed with respect to their differential expression and affinity towards the disease. The cancer specific target genes obtained in this approach were observed with significant prediction accuracy, which directs their use in prognostic application in diagnosis and treatment. These handpicked miRNA-target genes may further be biologically validated to confirm their role in biological processed and oncogenesis pathways.

References

1. Dikshit, R., et al.: Cancer mortality in India: a nationally representative survey. Lancet **379**(9828), 1807–1816 (2012)
2. Borse, V., Konwar, A.N., Buragohain, P.: Oral cancer diagnosis and perspectives in India. Sens. Int. **1**, 100046 (2020). https://doi.org/10.1016/j.sintl.2020.100046
3. Veluthattil, A.C., Sudha, S.P., Kandasamy, S., Chakkalakkoombil, S.V.: Effect of hypofractionated, palliative radiotherapy on quality of life in late-stage oral cavity cancer: a prospective clinical trial. Indian J. Palliat. Care **25**(3), 383 (2019)
4. Lucenteforte, E., Garavello, W., Bosetti, C., La Vecchia, C.: Dietary factors and oral and pharyngeal cancer risk. Oral Oncol. **45**(6), 461–467 (2009)
5. Fearon, E.R., Vogelstein, B.: A genetic model for colorectal tumorigenesis. Cell **61**(5), 759–767 (1990)
6. Bartel, D.P.: MicroRNAs: genomics, biogenesis, mechanism, and function. Cell **116**(2), 281–297 (2004)
7. Jansson, M.D., Lund, A.H.: MicroRNA and cancer. Mol. Oncol. **6**(6), 590–610 (2012)
8. Enerly, E., et al.: Correction: miRNA-mRNA integrated analysis reveals roles for miRNAs in primary breast tumors. PloS one **8**(9), e16915 (2013)
9. Lim, L.P., et al.: Microarray analysis shows that some microRNAs downregulate large numbers of target mRNAs. Nature **433**(7027), 769–773 (2005)
10. Seo, J., Jin, D., Choi, C.H., Lee, H.: Integration of microRNA, mRNA, and protein expression data for the identification of cancer-related microRNAs. PLoS One **12**(1), e0168412 (2017)
11. Bhowmick, S.S., Bhattacharjee, D., Rato, L.: Integrated analysis of the miRNA-mRNA next-generation sequencing data for finding their associations in different cancer types. Comput. Biol. Chem. **84**, 107152 (2020)
12. Sathipati, S.Y., Ho, S.Y.: Novel miRNA signature for predicting the stage of hepatocellular carcinoma. Sci. Rep. **10**(1), 1–12 (2020)
13. Varghese, R.S., et al.: Identification of miRNA-mRNA associations in hepatocellular carcinoma using hierarchical integrative model. BMC Med. Genom. **13**(1), 1–14 (2020)
14. Sarkar, J.P., Saha, I., Sarkar, A., Maulik, U.: Machine learning integrated ensemble of feature selection methods followed by survival analysis for predicting breast cancer subtype specific miRNA biomarkers. Comput. Biol. Med. **131**, 104244 (2021)

15. Falzone, L., et al.: Identification of novel MicroRNAs and their diagnostic and prognostic significance in oral cancer. Cancers **11**(5), 610 (2019)

16. Fang, C., Li, Y.: Prospective applications of microRNAs in oral cancer. Oncol. Lett. **18**(4), 3974–3984 (2019)

17. Tomczak, K., Czerwińska, P., Wiznerowicz, M.: The Cancer Genome Atlas (TCGA): an immeasurable source of knowledge. Contemp. Oncol. **19**(1A), A68 (2015)

18. Mahapatra, S., Mandal, B., Swarnkar, T.: Biological networks integration based on dense module identification for gene prioritization from microarray data. Gene Rep. **12**, 276–288 (2018)

19. Agarwal, V., Bell, G.W., Nam, J.W., Bartel, D.P.: Predicting effective microRNA target sites in mammalian mRNAs. Elife **4**, e05005 (2015)

20. Xiong, P., Schneider, R.F., Hulsey, C.D., et al.: Conservation and novelty in the microRNA genomic landscape of hyperdiverse cichlid fishes. Sci. Rep. **9**, 13848 (2019). https://doi.org/10.1038/s41598-019-50124-0

21. Hur, J.H., Ihm, S.Y., Park, Y.H.: A variable impacts measurement in random forest for mobile cloud computing. Wirel. Commun. Mob. Comput. (2017)

22. Mahapatra, S., Bhuyan, R., Das, J., Swarnkar, T.: Integrated multiplex network based approach for hub gene identification in oral cancer. Heliyon **7**(7), e07418 (2021)

23. Ardekani, A.M., Naeini, M.M.: The role of MicroRNAs in human diseases. Avicenna J. Med. Biotechnol. **2**(4), 161–79 (2010)

24. Han, H., Guo, X., Yu, H.: Variable selection using mean decrease accuracy and mean decrease gini based on random forest. In: 2016 7th IEEE International Conference on Software Engineering and Service Science (ICSESS). IEEE (2016)

Compact Associative Classification for Up and Down Regulated Genes Using Supervised Discretization and Clustering

S. Alagukumar[1], T. Kathirvalavakumar[2(✉)], and Rajendra Prasath[3]

[1] Department of Computer Applications, Ayya Nadar Janaki Ammal College, Sivakasi, Madurai Kamaraj University, Madurai, Tamil Nadu 626124, India
[2] Research Centre in Computer Science, V.H.N. Senthikumara Nadar College, Virudhunagar, Madurai Kamaraj University, Madurai, Tamil Nadu 626001, India
kathirvalavakumar@vhnsnc.edu.in
[3] Computer Science and Engineering Group, Indian Institute of Information Technology, Sri City, Chittoor, Andhra Pradesh 517646, India
rajendra.prasath@iiits.in

Abstract. Gene Expression data are being collected from microarray. Several rule mining algorithms are used to generate rules on microarray datasets but it produces large number of rules. To produce effective class association rules, such rules have to be grouped and pruned for analysis. To solve this problem we propose a method to reduce the number of class association rules and clustering them, particularly for the colon microarray gene expression data. The main objective is classifying the colon gene expression data using class association rule. The proposed methodology identifies the up and down regulated genes, generating class association rules, pruning class association rules and clustering the class association rules. Aim of the proposed clustering of class association rules is for identifying the highly correlated genes. Experimental results show that the proposed method gives best accuracy for the colon cancer dataset.

Keywords: Microarray gene expression · Differentially expressed genes · Class association rules · Clustering · Classification

1 Introduction

Colorectal cancer starts in the large intestine or rectum. One in 23 men and one in 25 women are affected by colon cancer during their lifetime. Colorectal cancer (CRC) accounts for 10% of yearly global cancer deaths [1]. Thus there is a need for efficient detection, treatment and identifying the most significant responsible genes for colon cancer. The microarray genes expression dataset contains large number of genes but small in sample size [20]. Many data mining techniques are used for analyzing the genes. The Association rule mining [1, 2] algorithm is one of the important data mining technique proposed by Agrawal [3] to solve market basket analysis. Association rule mining is used to solve classification and clustering problem. Classification is another

© Springer Nature Switzerland AG 2022
R. Chbeir et al. (Eds.): MIKE 2021, LNAI 13119, pp. 33–46, 2022.
https://doi.org/10.1007/978-3-031-21517-9_4

important task of data mining that can be used to analyze the training datasets to form accurate models. An associative classification performs important role to analyze dense data. An associative classification is an integration of association rules and classification. There are many classification algorithms called associative classification methods, which are using association rules. They are classification based association, classification based on multiple association rules and classification based on predicted association rules [6]. The associative classification algorithms produce higher accuracy than the tree-based classification algorithms. Moreover, Associative classification is also used to detect the most significant genes from the complex dataset. Here we propose a compact associative classificaton method for identifying the high correlated genes using class association rules and cluster the class association rules for microarray colon gene expression data. The paper is organized as follows. Section 2 highlights the review of the related work. Proposed method is presented in Sect. 3. Section 4 describes the experimental results and discussion. Conclusions are given in Sect. 5.

2   Literature Review

The differentially expressed genes and associative classification techniques are used to detect the most significant genes from the dataset and also enhance the classification accuracy [2, 4]. Ramesh et al. [3] have identified the differential gene expression on SARS – CoV infected samples using statistical test and log2 fold change. The performance of the associative classification can be increased by discretizing numerical attributes. Lavangnananda et al. [5] have reviewed the unsupervised discretization and supervised discretization methods on various classifiers. They have stated that the supervised discretization provides best performance than unsupervised discretization. Traditional Association rules are used to identify the relationship between the itemset and classification is used to predict the class label. However, Associative classification is used to find the relationship between the gene items and predict the class label. Thanajiranthorn et al. [6] have presented an associative classification to generate the effective rules without pruning process. Mattiev et al. [7] have proposed a method to cluster the class association rules using agglomerative algorithm, it reduces the number of class rules. Danh et al. [8] have analyzed the clustering of association rules and discovers the unexpected patterns. Rajab [9] has built a classifier based on rule-pruning called active rule pruning to improve predictive accuracy and to reduce rule redundancy. Mattiev and Kavsek [10] have stated that the accurate and compact classification is one of the vital task in data mining and over fitting prevention is an important task in an associative classification. They have proposed an associative classifier that selects strong class association rules based on coverage of the learning set. Azmi et al. [11] have pointed out the advantage of the associative classifiers and introduced an algorithm to build a classifier based on regularized class association rules in a categorical data space. This algorithm mines a set of class association rules based on support and confidence then apply a regularized logistic regression algorithm with Lasso penalty on the rules to build a classifier model [11]. Existing methods are used to diagnose and classify the gene expression using traditional algorithms. The proposed system uses the gene expression data of colon cancer patients because gene expression is responsible for causing any type of cancer. In the proposed method, the informative and up regulated and down regulated genes are identified from statistical test and log2 fold change and then Informative genes are fed into associative classification for classifying genes expressions as control or disease. Finally the gene expressions are grouped for indentifying the highly correlated patterns using clustering of the class association rules.

3 Proposed Method

The main motivation of the proposed associative classification model is to classify the genes as control or diseased genes and clustering the class association rules to identify highly correlated genes. The proposed method has five stages, i. Data Normalization, ii. Selecting the up and down regulated genes, iii. Supervised discretization, iv. Associative Classification and v. Clustering the class association rules.

3.1 Data Normalization

Before analyzing the data, the gene expression data must be normalized to avoid large variation in the gene expressions and to avoid errors in data processing [12]. The Z-score [2, 12] provides a way for standardizing data across a wide range of experiments and provides significant changes in the gene expression between different samples and conditions. Z score data normalization formula is given as follows:

$$Z = \frac{D - \mu}{\sigma} \tag{1}$$

where, D is the data to be normalized, μ is the arithmetic mean and σ is the standard deviation of that data and Z is the standardized variable with mean 0 and variance 1. This method is used to normalize the gene expression data.

3.2 Selection of Up and Down Regulated Genes

The differentially expressed genes (DEGs) of control and infected patients are identified using LIMMA test and log2 fold change value [3]. The multiple adjustment test [24] is used to calculate adjusted P-value. The differentially expressed genes are classified into up-regulated and down regulated genes using the log2 fold change value [4]. The differentially expressed genes (DEGs) are the subset of the significant genes which are used to improve the classification performance of the model.

3.3 Discretization

Discretization is a vital process in data mining to transform the continuous data into discretized data. The discretization algorithms can be classified as supervised and unsupervised [5]. The unsupervised discretization methods are having equal-width, equal-frequency and cluster based discretization. In the unsupervised methods, continuous features are divided into sub ranges by the user specified values. The unsupervised discretization does not give better accuracy for classification [13, 25, 26]. The supervised discretization methods find proper intervals with the cut-points using class information. The supervised discretization produces effective results. In this work, a Class-Attribute Contingency Coefficient (CACC) [13] based supervised discretization algorithm is used to transform the continuous gene expression into discrete data. The algorithm uses the class label for discretizing the data. The gene expression values are replaced with number of intervals using the discretization technique. CACC is the contingency coefficient value that is calculated between class variable and discrete intervals using the formula (2).

$$CACC = \sqrt{\frac{y}{y+S}} \qquad (2)$$

where $y = \chi^2/\log(n)$, S is the total number of attributes, n is the number of discretized intervals and y is the chi-square statistic.

3.4 Associative Classifier

The associative classification CBA is proposed by Liu et al. [14]. The class association rules are used for classification. A class association rule is an association rule where the antecedent of the rules contains gene items and consequent of the rule is a class label for the classification problem. The associative classification performs three steps [15, 16] namely, i. Generate the class association rules (CARs), ii. Prune the redundant rule, iii. Build the classifier model and predict the new data.

3.4.1 Generating Class Association Rules

The discretized data are transformed into transaction dataset. A transaction is a set of one or more genes extracted from up and down regulated gene expression data. The transaction data are used to generate the class association rules in two steps. First, all frequent item sets are found from the training dataset using minimum support. Then, from these frequent item sets class association rules are generated using minimum confidence. In the class association rules, the antecedent of the rule contains genes attributes and consequent of the rule contains only the class label. In generating class association rules, the Apriori Algorithm [17, 18] is used to find the frequent genes. A subset of a frequent item set must also be a frequent item set and if there is any infrequent item set then the infrequent item sets are to be removed. Iteratively all frequent item sets with cardinality 1 to k-item sets are found where k is the longest frequent gene item sets. After all the frequent item sets are generated from the training gene expression data, the class association rules are generated based on minimum support and minimum confidence. The confidence is calculated using the Eq. 3.

$$Confidence(A \rightarrow C) = \frac{Support_{Count}(A \cup C)}{Support_{Count}(A)} \qquad (3)$$

where A is an antecedent, C is a consequence, $Support_{Count}(A \cup C)$ is the number of transactions containing the itemsets $A \cup C$, and $Support_{Count}(A)$ is the number of transactions containing the itemsets A.

3.4.2 Pruning Redundant Rules

The class association rule algorithms for the gene expression data produces more rules and are hardly to understand [20, 27]. Hence these large numbers of rules are to be pruned without affecting its classifying performance. The class association rules are called as strong when it satisfies both the minimum support and minimum confidence. Here we generate strong class association rules. The class associations rules are pruned using constrains [19] and is stated as A → C is redundant if, A → C ∃ A' ⊂ A and confident (A' → C) ≥ confident(A → C).

3.4.3 Build the Classifier Model

The generated strong class association rules are the associative classifier model [14, 16, 20]. The class association rule is denoted as A → C, where A is the antecedent which contains gene items and C is the consequent which contains class label.

The test data are passed into the classification model to classify the gene expression and assess the performance of the model.

3.5 Clustering Class Association Rules

Interesting pattern finding is an important topic in data mining. An Associative classification is useful for analyzing the dense data. However, a drawback of mining is its large size of class association rules. Hence, the clustering of class association rules [21] is used to group the highly correlated rules and help to find the interesting pattern. Lent et al. [21] have introduced the concepts of clustering association rules. In this paper, Hierarchical clustering technique is used to group the class association rules. Clustering groups a set of n class association rules {CAR = CAR_1, CAR_2 . . . . . CAR_n} into m clusters {C = C_1, C_2 . . . . . C_m}. Let X_i and X_j are set of all items contained in CAR_i and CAR_j respectively where i, j ∈ {1, 2, . . . n} The jaccard similarity index is used to calculate the similarity between the class association rules and create the distance matrix using the Eq. (4).

$$\text{similarity}_{\text{jaccard}}\{X_i, X_j\} = 1 - \frac{|X_i \cap X_j|}{|X_i \cup X_j|} \tag{4}$$

In this paper, the single linkage hierarchical clustering method [22] is used to identify the distance between two clusters. The single linkage method identifies the minimum distance of two cluster objects using the Eq. (5) and finally forms a single cluster.

$$\text{SingleLinkage}(a, b) = \text{minimum}(D(x_{ai}, x_{bj})) \tag{5}$$

where x_{ai}, x_{bj} are two clusters in the sets i, j.

3.6 Pseudo Algorithm

Step 1: Read the Gene Expression Data
Step 2: Normalize the gene expression data

Step 3: Select the UP/Down regulated genes using LIMMA test
Step 4: Convert the continuous gene expression into discrete data using (2)
Step 5: Generate the class association rules
Step 6: Prune the redundant rules
Step 7: Build the Associative Classifier and Cluster the Class rules
Step 8: Predict the test data and correlate the data

4 Results and Discussion

The colon cancer-infected patients' gene expression data are collected from Kent ridge biomedical data repository, USA [2]. The colon gene expression dataset contains 2000 genes and 62 patients, where 40 patients are affected by cancer and 22 patients are control. The data of the microarray are presented in the gene expression matrix [20]. Experiments for the proposed method are carried out by R tool [23]. The gene expression matrix is normalized using the z-score. Among these normalized gene expression, 11 genes attributes are selected as the differentially expressed genes, where one gene attribute is up regulated genes and 10 gene attributes are down regulated genes. Table 1 represents the microarray colon cancer data set. Table 2 represents gene sequence id, expressed sequence tag (EST) number, gene accession number and gene description.

Table 1. Microarray colon gene expression data

Sample	Hsa.8147	Hsa.37937	Hsa.18321	Hsa.36689	Hsa.1131	Hsa.2456	Hsa.6814	Hsa.8125	Class
S1	−0.49	0.37	−0.03	−0.14	−0.43	−0.55	−0.22	−0.52	Cancer
S2	0.85	2.01	0.42	2.38	2.83	1.11	−0.68	0.70	Control
S3	−0.81	−0.28	−0.69	−0.48	−0.53	−0.84	−1.11	−1.24	Cancer
S4	−0.16	0.45	0.76	0.06	0.20	−0.66	−1.02	−1.21	Control
S5	−0.47	−0.71	−0.85	−1.08	−0.62	−0.48	0.32	−0.72	Cancer
S6	0.63	−0.16	−0.96	−0.59	−0.53	0.029	−0.64	0.55	Control
S7	−0.74	−0.71	−0.37	−0.60	−0.79	−0.65	−0.33	−0.83	Cancer
S8	2.06	0.31	−0.66	0.41	0.14	−0.84	−0.78	−0.39	Control

The differentially expressed genes (DEGs) between control and infected patients are identified with the help of LIMMA package in R programming [24]. The multiple adjustment test is used to calculate adjusted P-value. The differentially expressed genes are identified when its p value is <0.05 and log2 fold change value is >1.0. The colon gene expressions are classified into up-regulated and down-regulated genes using the log2 fold change (log2 FC) value. The description of the resultant differentially expressed genes are depicted in Table 3, where the cut off adjusted P-value is <0.05 and log2 fold change is >1.0 are used to find the differentially expressed genes. The fold change value greater than 1 indicates the up-regulated genes and others are indicated as down-regulated genes. It has been identified that 11 gene attributes are differentially expressed genes among the 2000 gene attributes.

Table 2. Colon cancer gene expression description

Sequence ID	Expressed sequence tag number	Gene accession number	Gene description
H55933	Hsa.3004	203417	mRNA for homologue
R39465	Hsa.13491	23933	Eukaryotic initiation factor 4A
R85482	Hsa.37254	180093	Serum response factor
R02593	Hsa.20836	124094	60s Acidic ribosomal protein P1
T51496	Hsa.1977	71488	60s Ribosomal protein l37A
H80240	Hsa.44472	240814	Inter-alpha-trypsin inhibitor complex component ii precursor
T65938	Hsa.3087	81639	Translationally controlled tumor protein
T55131	Hsa.1447	73931	Glyceraldehyde 3-phosphate dehydrogenase, liver
T72863	Hsa.750	84277	Ferritin light chain

Table 3. Selected top 11 differentially expressed genes

(Expressed sequence tag) EST number	logFC Value	Direction	Significant level of P-value
Hsa.8147	−1.309294	Down	0.001
Hsa.692.2	−1.236712	Down	0.001
Hsa.37937	−1.222846	Down	0.001
Hsa.1832	−1.219651	Down	0.001
Hsa.692	−1.209339	Down	0.001
Hsa.692.1	−1.192134	Down	0.001
Hsa.36689	−1.129647	Down	0.004
Hsa.1131	−1.121272	Down	0.004
Hsa.2456	−1.046896	Down	0.01
Hsa.6814	1.025599	Up	0.01
Hsa.8125	−1.01569	Down	0.02

The significant gene expressions are discretized into several intervals using CACC discretization algorithm. The discretized gene expressions are shown in the Table 4.

The discretized data are converted into the transactional dataset, which are used to generate the frequent item set and class association rules. The frequent items set are generated using the minimum support and the minimum confidence as 100%. Once the class association rules are generated the redundant rules are pruned. Finally the strong class association rules are generated, and are used to classify the test dataset. The class association rules are clustered to find the highly correlated genes using the single linkage hierarchical Clustering. Table 5 represents the class association rules and clustered index. Table 6 represents the comparative analysis of class association rules. The Compact associative classification is compared with other classifications namely [2], C50 Decision tree, Linear discriminant analysis and K-Nearest neighbor algorithms. Table 7 shows the comparison of different classification methods on colon gene expression data. The proposed method gives 100% accuracy.

Table 4. Discretized differentially expressed genes

Hsa.8147	Hsa.692.2	...	Hsa.2456	Hsa.6814	Hsa.8125	Class
[−0.843,0.399]	[−0.779,0.458]	...	[−0.816,0.659]	[−0.267,3.58]	[−1.41,0.416]	Cancer
[−0.843,0.399]	[−0.779,0.458]	...	[−0.816,0.659]	[−1.26,−0.267]	[−1.41,0.416]	Cancer
...	...	...	...	...	...	...
[−0.843,0.399]	[−0.779,0.458]	...	[−0.816,0.659]	[−0.267,3.58]	[0.416,2.4]	Cancer
[0.399,3.55]	[0.458,3.08]	...	[−0.816,0.659]	[−1.26,−0.267]	[0.416,2.4]	Control
[0.399,3.55]	[0.458,3.08]	...	[−0.927,−0.816]	[−1.26,−0.267]	[−1.41,0.416]	Control

Table 5. Generated class association rules

Rule No	LHS	RHS	Support in %	Confidence in %	Clustered Index
1	Hsa.692.2 = [0.47,4.34]	Class = control	100	100	1
2	Hsa.36689 = [−1.24,−0.6]	Class = cancer	100	100	2
3	Hsa.18321 = [0.216,1.37], Hsa.36689 = [0.655,3.22]	Class = control	100	100	1
4	Hsa.36689 = [0.655,3.22], Hsa.6814 = [−1.26,−0.267]	Class = control	100	100	1

(*continued*)

Table 5. (*continued*)

Rule No	LHS	RHS	Support in %	Confidence in %	Clustered Index
5	Hsa.18321 = [0.216,1.37], Hsa.2456 = [0.659,2.94]	Class = control	100	100	1
6	Hsa.18321 = [0.216,1.37], Hsa.692 = [0.433,2.85]	Class = control	100	100	1
7	Hsa.18321 = [0.216,1.37], Hsa.692.1 = [0.521,2.67]	Class = control	100	100	1
8	Hsa.8147 = [0.379,3.55], Hsa.18321 = [0.216,1.37]	Class = control	100	100	1
9	Hsa.36689 = [−0.6,0.655], Hsa.6814 = [−0.267,3.58]	Class = cancer	100	100	2
10	Hsa.18321 = [−1.36,0.216], Hsa.6814 = [−0.267,3.58]	Class = cancer	100	100	2
11	Hsa.37937 = [−0.917,−0.188], Hsa.6814 = [−0.267,3.58]	Class = cancer	100	100	2
12	Hsa.6814 = [−0.267,3.58], Hsa.8125 = [−1.41,0.403]	Class = cancer	100	100	2
13	Hsa.1131 = [−0.883,0.208], Hsa.6814 = [−0.267,3.58]	Class = cancer	100	100	2

(*continued*)

Table 5. (*continued*)

Rule No	LHS	RHS	Support in %	Confidence in %	Clustered Index
14	Hsa.8147 = [−0.843,0.379], Hsa.6814 = [−0.267,3.58]	Class = cancer	100	100	2
15	Hsa.692 = [−0.963,0.433], Hsa.6814 = [−0.267,3.58]	Class = cancer	100	100	2
16	Hsa.2456 = [−0.949,0.659], Hsa.6814 = [−0.267,3.58]	Class = cancer	100	100	2
17	Hsa.692.1 = [−0.995,0.521], Hsa.6814 = [−0.267,3.58]	Class = cancer	100	100	2
18	Hsa.37937 = [−0.188,3.89], Hsa.2456 = [0.659,2.94], Hsa.6814 = [−1.26,−0.267]	Class = control	100	100	1
19	Hsa.37937 = [−0.188,3.89], Hsa.692 = [0.433,2.85], Hsa.6814 = [−1.26,−0.267]	Class = control	100	100	1
20	Hsa.37937 = [−0.188,3.89], Hsa.692.1 = [0.521,2.67], Hsa.6814 = [−1.26,−0.267]	Class = control	100	100	1

(*continued*)

Table 5. (*continued*)

Rule No	LHS	RHS	Support in %	Confidence in %	Clustered Index
21	Hsa.8147 = [0.379,3.55], Hsa.37937 = [−0.188,3.89], Hsa.6814 = [−1.26,−0.267]	Class = control	100	100	1
22	Hsa.18321 = [0.216,1.37], Hsa.1131 = [0.208,3.06], Hsa.8125 = [0.403,2.4]	Class = control	100	100	1
23	Hsa.18321 = [0.216,1.37], Hsa.6814 = [−1.26,−0.267], Hsa.8125 = [0.403,2.4]	Class = control	100	100	1
24	Hsa.37937 = [−0.188,3.89], Hsa.6814 = [−1.26,−0.267], Hsa.8125 = [0.403,2.4]	Class = control	100	100	1

Table 6. Comparison of class association rules for different parameters

Support in %	Confidence in %	Before rules prune	After rules prune
20	100	1536	65
40	100	1115	46
60	100	833	32
80	100	719	26
100	100	610	24

Figure 1 indicates the single linkage hierarchical clustering tree. It has unique splitting points and each result in a different cluster set. For example splitting at height of 1.0 gives three cluster sets for control gene groups (Rule 1), (Rule 3, Rule 4, Rule 5, Rule 6, Rule 7 and Rule 8) and (Rule 18, Rule 19, Rule 20, Rule 21, Rule 22, Rule 23, Rule 24) and two cluster sets for cancer gene groups (Rule 2), (Rule 9, Rule 10, Rule

Table 7. Comparison of classification techniques

Methods	Accuracy in %	Error rate in %
Wilcoxon sign rank sum test and random forest [2]	99.81	0.19
Proposed CAC model	100	0.0
C50	98	2.0
LDA	92.5	7.5
KNN	98	2.0

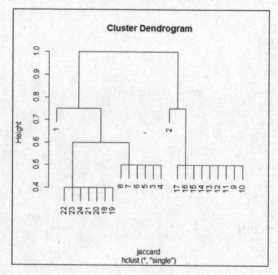

Fig. 1. Single linkage hierarchical clustering of class association rules

11, Rule 12, Rule 13, Rule 14, Rule 15, Rule 16, Rule 17). Here rules in each cluster are highly correlated genes.

5   Conclusion and Future Work

Microarray data are producing more rules when classifying gene expression data. By grouping and pruning the class association rules effective class association rules can be generated. The proposed method reduces the number of class association rules and clustered the class association rules on colon microarray gene expression data. The proposed methodology identifies the up and down regulated genes using LIMMA test and log2 fold change, generates class association rules, prune the class association rules and identifies the highly correlated rules among the CARs using single linkage hierarchical clustering. The experimental results performed on colon cancer datasets gives 100% accuracy. In the experiment normal frequent itemsets are used to generate class association rules but this leads to generate more rules. In future this can be minimized by using different types of frequent itemsets instead of normal frequent itemset. Instead of only using colon cancer datasets various cancer datasets can be used in the experiment.

References

1. Thomas, V.M., et al.: Trends in colorectal cancer incidence in India. J. Clin. Oncol. **38**(15_suppl), e16084–e16084 (2020). https://doi.org/10.1200/JCO.2020.38.15_suppl.e16084
2. Maniruzzaman, M., et al.: Statistical characterization and classification of colon microarray gene expression data using multiple machine learning paradigms. Comput. Methods Programs Biomed. **176**, 173–193 (2019)
3. Ramesh, P., Veerappapillai, S., Karuppasamy, R.: Gene expression profiling of corona virus microarray datasets to identify crucial targets in COVID-19 patients. Gene Rep. **22** (2021)
4. Dingerdissen, H.M., Vora, J., Cauley, E., Bell, A., King, C.H., Mazumder, R.: Differential expression of glycosyltransferases identified through comprehensive pan-cancer analysis (2021)
5. Lavangnananda, K., Chattanachot, S.: Study of discretization methods in classification. In: 2017 9th International Conference on Knowledge and Smart Technology (KST), pp. 50–55. IEEE (2017)
6. Thanajiranthorn, C., Songram, P.: Efficient rule generation for associative classification. Algorithms **13**(11), 299 (2020)
7. Mattiev, J., Kavšek, B.: CMAC: clustering class association rules to form a compact and meaningful associative classifier. In: Nicosia, G., et al. (eds.) LOD 2020. LNCS, vol. 12565, pp. 372–384. Springer, Cham (2020). https://doi.org/10.1007/978-3-030-64583-0_34
8. Bui-Thi, D., Meysman, P., Laukens, K.: Clustering association rules to build beliefs and discover unexpected patterns. Appl. Intell. **50**(6), 1943–1954 (2020). https://doi.org/10.1007/s10489-020-01651-1
9. Rajab, K.D.: New associative classification method based on rule pruning for classification of datasets. IEEE Access **7**, 157783–157795 (2019)
10. Mattiev, J., Kavsek, B.: Coverage-based classification using association rule mining. Appl. Sci. **10**(20), 7013 (2020)
11. Azmi, M., Runger, G.C., Berrado, A.: Interpretable regularized class association rules algorithm for classification in a categorical data space. Inf. Sci. **483**, 313–331 (2019)
12. Cheadle, C., Vawter, M.P., Freed, W.J., Becker, K.G.: Analysis of microarray data using Z score transformation. J. Mol. Diagn. **5**(2), 73–81 (2003)
13. Tsai, C.J., Lee, C.I., Yang, W.P.: A discretization algorithm based on class-attribute contingency coefficient. Inf. Sci. **178**(3), 714–731 (2008)
14. Liu, B., Hsu, W., Ma, Y.: Integrating classification and association rule mining. In: Proceedings of the Fourth International Conference on Knowledge Discovery and Data Mining, KDD'98, pp. 80–86. AAAI Press (1998)
15. Vanhoof, K., Depaire, B.: Structure of association rule classifiers: a review. In: International Conference on Intelligent Systems and Knowledge Engineering (ISKE), pp. 9–12 (2010)
16. Hahsler, M., Johnson, I., Kliegr, T., Kucha, J.: Associative classification in R: arc, arulesCBA, and rCBA. R Journal **9**(2) (2019)
17. Agrawal, R., Imielinski, T., Swami, A.: Mining association rules between sets of items in large databases. In: Proceedings of the ACM SIGMOD International Conference on Management of Data, pp. 207–216. ACM Press (1993)

18. Agrawal, R., Srikant, R.: Fast algorithms for mining association rules in large databases. In: Proceedings of the 20th International Conference on Very Large Data Bases, VLDB '94, pp. 487–499. Morgan Kaufmann Publishers Inc., San Francisco, CA, USA (1994). ISBN 1-55860-153-8

19. Bayardo, R.J., Agrawal, R., Gunopulos, D.: Constraint-based rule mining in large, dense databases. Data Min. Knowl. Disc. 4(2), 217–240 (2000)

20. Alagukumar, S., Lawrance, R.: Classification of microarray gene expression data using associative classification. In: IEEE International Conference on Computing Technologies and Intelligent Data Engineering (ICCTIDE), pp. 1–8 (2016)

21. Lent, B., Swami, A., Widom, J.: Clustering association rules. In: 13-th IEEE International Conference on Data Engineering, pp. 220–23 (1997)

22. Nielsen, F.: Hierarchical clustering. In: Introduction to HPC with MPI for Data Science, pp. 195–211. Springer (2016)

23. https://www.r-project.org/

24. Ritchie, M.E., et al.: limma powers differential expression analyses for RNA-sequencing and microarray studies. Nucleic Acids Res. 43(7) (2015)

25. Liu, H., Hussain, F., Tan, C.L., Dash, M.: Discretization: an enabling technique. Data Min. Knowl. Disc. 6(4), 393–423 (2002)

26. Hacibeyoğlu, M., Ibrahim, M.H.: Comparison of the effect of unsupervised and supervised discretization methods on classification process. Int. J. Intell. Syst. Appl. Eng. 105–108 (2016)

27. Abdelhamid, N., Thabtah, F.: Associative classification approaches: review and comparison. J. Inf. Knowl. Manag. 13(03) (2014)

Assessment of Brain Tumor in Flair MRI Slice with Joint Thresholding and Segmentation

Seifedine Kadry[1,2,3](✉) , David Taniar[4] , Maytham N. Meqdad[5],
Gautam Srivastava[6] , and Venkatesan Rajinikanth[7]

[1] Department of Applied Data Science, Noroff University College, Kristiansand, Norway
skadry@gmail.com
[2] Artificial Intelligence Research Center (AIRC), College of Engineering and Information
Technology, Ajman University, Ajman, United Arab Emirates
[3] Department of Electrical and Computer Engineering, Lebanese American University,
Byblos, Lebanon
[4] Faculty of Information Technology, Monash University, Monash, Australia
[5] Al-Mustaqbal University College, Hillah, Babil, Iraq
[6] Department of Mathematics and Computer Science, Brandon University, 270 18th Street,
Brandon R7A 6A9, Canada
[7] Department of Electronics and Instrumentation, St. Joseph's College of Engineering, Chennai,
Tamilnadu 600 119, India

Abstract. Medical image assessment plays a vital role in hospitals during the disease assessment and decision making. Proposed work aims to develop an image processing procedure to appraise the brain tumor fragment from Flair modality recorded MRI slice. The proposed technique employs joint thresholding and segmentation practice to extract the infected part from the chosen image. Initially, a tri-level thresholding based on Mayfly Algorithm and Kapur's Entropy (MA + KE) is implemented to improve the tumor and then the tumor area is mined using the automated Watershed Segmentation Scheme (WSS). The merit of the employed procedure is verified on various 2D MRI planes, such as axial, coronal and sagittal and the experimental outcome confirmed that this technique helps to mine the tumor area with better accuracy. In this work, the necessary images are collected from BRATS2015 dataset and 30 patient's information (10 slices per patient) is considered for the examination. The experimental investigation is implemented using MATLAB® and 300 images from every 2D plane are examined. The proposed technique helps to get better values of Jaccard-Index (>85%), Dice-coefficient (>91%) and Accuracy (98%) on the considered MRI slices.

Keywords: Brain tumor · Flair modality · Mayfly algorithm · Kapur · Watershed algorithm

1 Introduction

Brain Tumor (BT) is one of the harsh illnesses in human group and the timely recognition and treatment is essential to save the patient. The untreated BT will cause various

© Springer Nature Switzerland AG 2022
R. Chbeir et al. (Eds.): MIKE 2021, LNAI 13119, pp. 47–56, 2022.
https://doi.org/10.1007/978-3-031-21517-9_5

problems and timely recognition and decision making is necessary to execute the recommended treatment procedure [1–3]. Most of the brain tumor is recognized only after getting a severe disease symptom and identification of the tumor, its location in brain and its dimension is a prime task to suggest the necessary handling procedure and medication.

Due to its significance, a number of traditional and soft-computing supported BT detection procedures are proposed and executed on the benchmark and clinically collected MRI slices [4, 5]. Most of the clinical brain MRI is associated with the skull-region and hence, a skull stripping methods are also employed to reduce the disease detection complexity [6, 7]. The literature confirms that, the MRI can be recorded in a controlled environment under the guidance of an experienced radiologist and based on the suggestion and patients, health condition, the contrast agents are employed/ignored during the scanning of brain sections. After recording the brain region with MRI of chosen modality, the radiologist will perform the initial analysis on the image and the evaluation report along with the reconstructed 3D image is then submitted for the examination of an experienced brain specialist. The doctor will perform an assessment on the MRI and based on the doctor's observation and the report supplied by a radiologist, the decision regarding the disease and its treatment is arrived. The normal treatment involves in surgery followed by radiotherapy. When the treatment for BT is employed; the recovery rate also inspected by recording the MRI.

The segmentation of BT can be achieved with traditional segmentation and Convolutional-Neural-Network (CNN) schemes and the implementation of traditional technique is quite easy compared to CNN methods and hence a number of joint thresholding schemes are implemented by the researchers. In this technique, the tumor region is to be improved with a chosen thresholding approach and then the BT in improved image is mined [8, 9]. This approach helps to achieve good result on the considered brain MRI slices and its outcome can be used to improved the decision making process.

In the proposed research, an image examination technique is implemented with the help of joint thresholding and segmentation and the initial section of this work consist a tri-level thresholding via Mayfly Algorithm and Kapur Entropy (MA + KE) and the enhanced tumor part of thresolded image is then segmented with a watershed algorithm. This technique is an automated approach helps to extract the BT with better segmentation technique.

The proposed work considered 30 patient's images collected from the BRATS2015 for the assessment. It is a 3D image database and 3D to 2D alteration is achieved with ITK-Snap [10, 11] and the extracted images of axial, coronal and sagittal planes are individually examined and mined tumor region is compared with the Ground-Truth (GT) for the confirmation. In this work, 300 images of each plane is separately examined and for every 2D plane, a better result is achieved. Watershed Segmentation Scheme (WSS) is an easy and automated approach, helps to extract the tumor section with better segmentation accuracy. All the investigation is executed with MATLAB® and the achieved result of this scheme is agreeable and provided better segmentation results. In future, this technique can be tested and verified using the clinically collected brain MRI slices.

The contribution of this work comprises:

- Development of a simple automated BT segmentation using MA + KE and WSS.
- Testing and validating the performance on various MRI planes

Other parts of the work arranged as; Sect. 2 shows earlier research work, Sect. 3 presets the methodology, Sects. 4 and 5 demonstrate investigational result and the conclusion respectively.

2 Related Research

Automatic examination of brain MRI slice plays a vital function in brain illness detection procedures. The irregularity in brain is generally recorded by different modalities based on the need and recorded image is assessed with a chosen imaging scheme. Table 1 presents different traditional examination methods existing for the brain MRI slices.

Table 1. Summary of brain tumour examination with MRI slices

Reference	Examination technique employed
Fernandes et al. [1]	Implementation of a consistent structure for early diagnosis of brain tumor detection and treatment planning is implemented and tested with clinically collected MRI slices
Roopini et al. [2]	Joint thresholding and segmentation of BT using fuzzy entropy and level set.is presented and discussed using the images collected from BRATS2015
Rajinikanth et al. [3]	VGG19 supported segmentation and classification of brain MRI is presented and in this research benchmark and clinical images are examined
Rajinikanth et al. [4]	CNN segmentation and SVM classification is implemented to examine the brain tumor in benchmark and clinical images
Kadry et al. [5]	Implementation of joint thresholding and segmentation is diacussed using BRATS2015 database. In this work, tri-level thresholding is implemented using Moth-Flame algorithm and KE and the segmentation is implemented with traditional procedures, including the WSS
Lin et al. [6]	Implementation of joint thresholding and segmentation procedure to examine the brain MRI of stroke/tumor is presented
Satapathy & Rajinikanth [7]	Jaya algorithm based tri-level thresholding and segmentation is presented. In this work entropy based thresholding is considered for the tumor enhancement and chosen segmentation is adopted to extract the BT
Manic et al. [8]	Implementation of KE based thresholding and Chan-Vese segmentation is performed to extract BT from MRI slices
Kadry et al. [9]	Implementation of CNN based segmentation and classification of BT and stroke is presented using BRATS2015 database

Table 1 demonstrates the existing methodologies to extract and evaluate the brain tumor from various MRI modalities. Evaluation of the tumor dimension and its location in brain is vital information for the doctor, who plan and execute the treatment. Accurate mining of tumor from MRI is a challenging task and automatic mining is always preferred. The proposed research implements a simple procedure to mine the BT. This process integrates the thresholding and segmentation technique to mine the BT with better accuracy.

3   Methodology

This section of the research presents the methodology employed to extract the tumor with better accuracy. The different phases of this work are clearly presented in Fig. 1. The initial work is the collection of 3D images from the BRATS2015 database. The 3D to 2D conversion is employed using ITK-Snap and this process helps to get different planes, like axial, coronal and Sagittal. After extracting the 2D slices from Flair modality MRI, its related GT also then collected and separately stored. The enhancement of the tumor region in MRI slice is then achieved with tri-level thresholding executed by Mayfly Algorithm and Kapur Entropy (MA + KE). After the enhancement, the BT is then mined using Watershed-Segmentation-Scheme (WSS) and finally, the mined BT is compared with the GT and performance of the system is confirmed based on attained performance values.

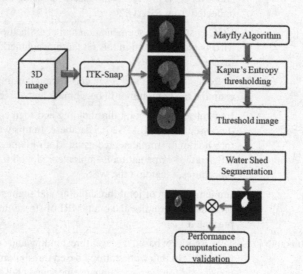

Fig. 1. Structure of joint thresholding and segmentation technique to mine brain tumor

Figure 1 depicts the various phases available in the proposed pipeline and the collected Whole Histology Slide (WHS) is divided into 25 patches and resized to a chosen dimension. The features from these images are separately extracted with DL and DWT approach and then MA is employed to select the optimal features. The selected features are then concatenated serially to get a new one dimensional (1D) feature vector (DF + HF) and these features are then considered to train and validate the classifiers using a 5-fold cross validation.

3.1 BRATS2015 Database

The benchmark medical images are extensively considered for testing and validating the performance of image inspection schemes. In this work, the necessary 2D slices are collected from BRATS2015 database [12, 13]. The High-Grade-Glioma (HGG) class BT is collected from 30 patient's images and from each patient 10 slices are extracted. All the 2D planes are considered for the assessment and from every case, 300 images of dimension $240 \times 240 \times 3$ pixels is assessed with the proposed technique. The obtained result is separately demonstrated for all these three 2D planes and this research helped to get better result irrespective of MRI plane.

Figure 2 presents the sample test image of various planes along with its GT and Fig. 3 presents the sample images of axial plane.

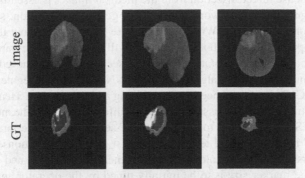

Fig. 2. Sample test images of different 2D planes

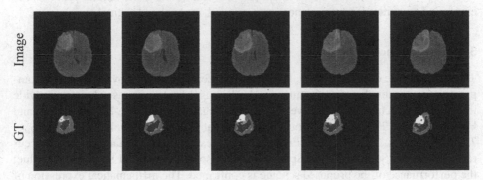

Fig. 3. Sample axial plane images of a chosen patient

3.2 Tri-Level Thresholding

The integration of image pre-processing and post-processing works are widely implemented in medical image processing works to get appropriate outcome. The earlier works in the literature confirmed the eminence of tri-level thresholding achieved with a chosen thresholding process along with a heuristic algorithm. In this work, the pre-processing of the brain MRI slice is achieved using Mayfly Algorithm and Kapur Entropy (MA + KE) and the enhanced tumor section is then mined with an automated segmentation procedure.

The recent thresholding work of Kadry et al. (2021) [14] discussed the implementation of MA and Otsu function to achieve thresholding process and in this work, the Otsu is replaced with KE and the enhancement of tumor is achieved with a chosen threshold level of three. The earlier work related to KE thresholding of brain MRI can be found in [5].

3.3 Automatic Segmentation

Extraction of disease infected section from the medical images is essential work during the disease diagnosis and this work will help to get the information, such as, infection rate, location and severity. When the complete information about the infection is available, appropriate decision about the disease will be made before implementing the treatment. Normally, the segmentation of disease section can be achieved with, manual operator, tradition segmentation algorithms and CNN. The manual segmentation is time consuming and CNN technique needs more training and testing time. Hence, traditional algorithms are still employed to assess the disease information from the medical images. In this work, WSS is employed to extract the BT segment from the thresholded MRI slice. The WSS is an automated technique, consist series of operations, such as edge detection, watershed fill, morphological enhancement, localization and segmentation. In WSS, only the marker size is to be initially tuned by the operator and based on its value, it will extract the section automatically. The earlier works on BT segmentation with WSS can be found in the literature [15–17] and in this work, similar procedure is employed to extract BT from planes, such as axial, coronal and sagittal.

3.4 Performance Evaluation

The performance of the automatic BT mining process is validated by employing a comparative assessment between the tumor and GT image. This comparison helps to provide initial values, such as True-Positive (TP), False-Negative (FN), True-Negative (TN), and False-Positive (FP) values and from these information, its rates, such as TP_{rate}, FN_{rate}, TN_{rate} and FP_{rate} are computed. Further, other measures, such as Jaccard, Dice, Accuracy, Precision, Sensitivity and Specificity are also computed and based on its value; the performance of the proposed scheme is confirmed. The mathematical expression of these parameters can be accessed from [18–22].

4 Result and Discussion

This part of the work demonstrates the experimental result and discussions. The proposed research is implemented using MATLAB® with a workstation; Intel i7 2.9 GHz processor, 20 GB RAM and 4 GB VRAM.

Initially, the necessary 2D slices of different planes are obtained using ITK-Snap and in these work 300 images from every category are examined with the proposed technique. Before the examination, every image is resized into $240 \times 240 \times 3$ pixels.

The proposed scheme consist the integration of MA + KE thresholding and WSS based mining. After extracting the BT, a pixel wise comparison is then executed with the GT and the necessary performance values are computed. Based on these values, the eminence of the proposed scheme is confirmed.

The proposed MA + KE based thresholding and WSS is implemented on the considered test images of different planes and the sample results are depicted in Fig. 4. Figure 4(a) depicts the test image and detected edges, Fig. 4(b) to (d) shows, watershed image, and morphological enhancement and enhanced tumor section. Figure 4(e) and (f) shows the extracted BT and its GT. Finally, the mined BT is compared with GT and the performance values are computed. Tables 2 and 3 depicts the performance values of sample images depicted in Fig. 4 and similar procedure is employed for other images considered in this study. These tables confirm that the results achieved with the proposed scheme are helping to achieve a better performance measures on all the considered 2D planes. Figure 5 compares the necessary measures achieved with axial, coronal and sagittal planes and in every class the measures obtained are superior.

Similar technique is then employed for all other image and the average value of the values (Mean ± Standard Deviation) is depicted in Table 4. This result also confirms that the proposed technique helps to get a better value of Jaccard, Dice and Accuracy on the considered images.

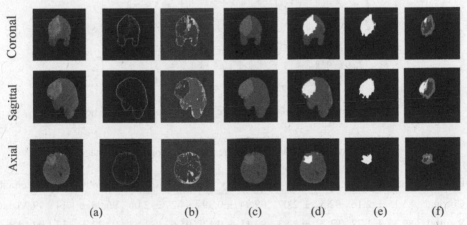

Fig. 4. Experimental outcome of the proposed scheme. (a) Edge detection, (b) Watershed fill, (c) Morphology, (d) Enhanced tumour, (e) Mined tumour, and (f) GT

In the proposed research, an effective automatic BT examination is proposed and implemented and in future, the performance of the joint thresholding and segmentation scheme can be verified with other approaches existing in the literature. Furthermore, the performance of this scheme can be tested and verified on; (i) brain MRI associated with the skull section and (ii) Clinically collected brain MRI slices.

Table 2. Initial result achieved with the proposed scheme

Plane	TP	FN	TN	FP	TP_{rate}	FN_{rate}	TN_{rate}	FP_{rate}
Coronal	4024	191	53102	283	0.9547	0.0453	0.9947	0.0053
Sagittal	4052	497	52997	54	0.8907	0.1093	0.9990	0.0010
Axial	2678	267	54627	19	0.9093	0.0907	0.9997	0.0003

Table 3. Essential performance measure computed for the sample test images

Plane	Jaccard	Dice	Accuracy	Precision	Sensitivity	Specificity
Coronal	89.4620	94.4379	99.1771	93.4293	95.4686	99.4699
Sagittal	88.0295	93.6337	99.0434	98.6849	89.0745	99.8982
Axial	90.3509	94.9309	99.5034	99.2955	90.9338	99.9652

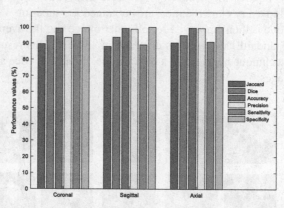

Fig. 5. Comparison of performance measures for various planes

Table 4. Average performance values achieved for 300 test images

Plane	Jaccard	Dice	Accuracy	Precision	Sensitivity	Specificity
Coronal	85.15 ± 2.16	92.52 ± 2.07	98.94 ± 0.39	94.27 ± 2.06	96.08 ± 1.14	99.26 ± 0.08
Sagittal	87.36 ± 1.72	91.31 ± 2.83	99.11 ± 0.17	95.63 ± 1.83	92.22 ± 3.13	99.14 ± 0.14
Axial	88.18 ± 2.06	91.52 ± 1.74	98.14 ± 0,62	97.52 ± 2.04	91.53 ± 2.95	99.03 ± 0.32

5   Conclusion

This research work implemented a joint thresholding and segmentation scheme to extract the BT from 2D MRI slice of different planes. In this work, the tumor section is initially enhanced using tri-level thresholding executed with MA + KE and then the tumor pat is mined using WSS. This work is an automated technique and helps to extract the tumor from the test image with better segmentation accuracy. After the extraction, the BT is compared against GT and the necessary measures are computed. The overall result of this work helps to get the following values of Jaccard (>85%), Dice (>91%) and Accuracy (98%) on the considered MRI slices and this confirms its merit in evaluating the BT section in the MRI slices of Flair modality. In future, the performance of the proposed scheme can be validated on; MRI slices with different modality, MRI slice with associated skull section and clinical grade MRI images collected from real patients.

References

1. Fernandes, S.L., Tanik, U.J., Rajinikanth, V., Karthik, K.A.: A reliable framework for accurate brain image examination and treatment planning based on early diagnosis support for clinicians. Neural Comput. Appl. **32**(20), 15897–15908 (2019). https://doi.org/10.1007/s00521-019-04369-5
2. Thivya Roopini, I., Vasanthi, M., Rajinikanth, V., Rekha, M., Sangeetha, M.: Segmentation of tumor from brain MRI using fuzzy entropy and distance regularised level set. In: Nandi, A.K., Sujatha, N., Menaka, R., Alex, J.S.R. (eds.) Computational Signal Processing and Analysis. LNEE, vol. 490, pp. 297–304. Springer, Singapore (2018). https://doi.org/10.1007/978-981-10-8354-9_27
3. Rajinikanth, V., Joseph Raj, A.N., Thanaraj, K.P., Naik, G.R.: A customized VGG19 network with concatenation of deep and handcrafted features for brain tumor detection. Appl. Sci. **10**(10), 3429 (2020)
4. Rajinikanth, V., Kadry, S., Nam, Y.: Convolutional-neural-network assisted segmentation and SVM classification of brain tumor in clinical MRI slices. Information Technology and Control **50**(2), 342–356 (2021)
5. Kadry, S., Rajinikanth, V., Raja, N.S.M., Jude Hemanth, D., Hannon, N.M.S., Raj, A.N.J.: Evaluation of brain tumor using brain MRI with modified-moth-flame algorithm and Kapur's thresholding: a study. Evol. Intel. **14**(2), 1053–1063 (2021). https://doi.org/10.1007/s12065-020-00539-w
6. Lin, D., Rajinikanth, V., Lin, H.: Hybrid image processing-based examination of 2D brain MRI slices to detect brain tumor/stroke section: a study. In: Priya, E., Rajinikanth, V. (eds.) Signal and Image Processing Techniques for the Development of Intelligent Healthcare Systems, pp. 29–49. Springer, Singapore (2021). https://doi.org/10.1007/978-981-15-6141-2_2
7. Satapathy, S.C., Rajinikanth, V.: Jaya algorithm guided procedure to segment tumor from brain MRI. J. Optim. **2018**, 1–12 (2018). https://doi.org/10.1155/2018/3738049
8. Suresh Manic, K., Hasoon, F.N., Shibli, N.A., Satapathy, S.C., Rajinikanth, V.: An approach to examine brain tumor based on Kapur's entropy and Chan–Vese algorithm. In: Yang, X.-S., Sherratt, S., Dey, N., Joshi, A. (eds.) Third International Congress on Information and Communication Technology. AISC, vol. 797, pp. 901–909. Springer, Singapore (2019). https://doi.org/10.1007/978-981-13-1165-9_81
9. Kadry, S., Nam, Y., Rauf, H.T., Rajinikanth, V., Lawal, I.A.: Automated detection of brain abnormality using deep-learning-scheme: a study. In: 2021 Seventh International Conference on Bio Signals, Images, and Instrumentation (ICBSII), pp. 1–5. IEEE (Mar 2021)
10. Yushkevich, P.A., Gao, Y., Gerig, G.: ITK-SNAP: an interactive tool for semi-automatic segmentation of multi-modality biomedical images. In: 2016 38th Annual International Conference of the IEEE Engineering in Medicine and Biology Society (EMBC), pp. 3342–3345. IEEE (Aug 2016)

11. ITK-Snap. http://www.itksnap.org/pmwiki/pmwiki.php
12. Menze, B.H., et al.: The multimodal brain tumor image segmentation benchmark (BRATS). IEEE Trans. Med. Imaging **34**(10), 1993–2024 (2014)
13. Menze, B.H., Van Leemput, K., Lashkari, D., Weber, M.A., Ayache, N., Golland, P.: A generative model for brain tumor segmentation in multi-modal images. In: International Conference on Medical Image Computing and Computer-Assisted Intervention, pp. 151–159. Springer, Berlin, Heidelberg (Sep 2010)
14. Kadry, S., Rajinikanth, V., Koo, J., Kang, B.-G.: Image multi-level-thresholding with Mayfly optimization. Int. J. Electr. Comput. Eng. (IJECE) **11**(6), 5420 (2021). https://doi.org/10.11591/ijece.v11i6.pp5420-5429
15. Levner, I., Zhang, H.: Classification-driven watershed segmentation. IEEE Trans. Image Process. **16**(5), 1437–1445 (2007)
16. Nguyen, H.T., Worring, M., Van Den Boomgaard, R.: Watersnakes: energy-driven watershed segmentation. IEEE Trans. Pattern Anal. Mach. Intell. **25**(3), 330–342 (2003)
17. Shafarenko, L., Petrou, M., Kittler, J.: Automatic watershed segmentation of randomly textured color images. IEEE Trans. Image Process. **6**(11), 1530–1544 (1997)
18. Khan, M.A., et al.: Computer-aided gastrointestinal diseases analysis from wireless capsule endoscopy: a framework of best features selection. IEEE Access **8**, 132850–132859 (2020)
19. Shree, et al.: A hybrid image processing approach to examine abnormality in retinal optic disc. Procedia Comput. Sci. **125**, 157–164 (2018). https://doi.org/10.1016/j.procs.2017.12.022
20. Dey, N., Rajinikanth, V., Fong, S.J., Kaiser, M.S., Mahmud, M.: Social group optimization–assisted Kapur's entropy and morphological segmentation for automated detection of COVID-19 infection from computed tomography images. Cogn. Comput. **12**(5), 1011–1023 (2020)
21. Rajinikanth, V., Thanaraj, K.P., Satapathy, S.C., Fernandes, S.L., Dey, N.: Shannon's entropy and watershed algorithm based technique to inspect ischemic stroke wound. In: Smart Intelligent Computing and Applications, pp. 23–31. Springer, Singapore (2019)
22. Fernandes, S.L., Rajinikanth, V., Kadry, S.: A hybrid framework to evaluate breast abnormality using infrared thermal images. IEEE Consum. Electron. Mag. **8**(5), 31–36 (2019)

Mayfly-Algorithm Selected Features for Classification of Breast Histology Images into Benign/Malignant Class

Seifedine Kadry[1,2,3]([✉]) [iD], Venkatesan Rajinikanth[4] [iD], Gautam Srivastava[5] [iD], and Maytham N. Meqdad[6]

[1] Department of Applied Data Science, Noroff University College, Kristiansand, Norway
skadry@gmail.com
[2] Artificial Intelligence Research Center (AIRC), College of Engineering and Information Technology, Ajman University, Ajman, United Arab Emirates
[3] Department of Electrical and Computer Engineering, Lebanese American University, Byblos, Lebanon
[4] Department of Electronics and Instrumentation, St. Joseph's College of Engineering, Chennai 600 119, Tamilnadu, India
[5] Department of Mathematics and Computer Science, Brandon University, 270 18th Street, Brandon R7A 6A9, Canada
[6] Al-Mustaqbal University College, Hillah, Babil, Iraq

Abstract. Incidence rate of Breast Cancer (BC) is rising globally and the early detection is important to cure the disease. The detection of BC consist different phases from verification to clinical level diagnosis. Confirmation of the cancer and its stage is performed normally with breast biopsy. This research aims to develop a framework to identify Benign/Malignant class images from the Breast Histology Slide (BHS). This technique consist the following phases; (i) Cropping and resizing the image slice, (ii) Deep-feature extraction using pre-trained network, (iii) Discrete Wavelet Transform (DWT) feature mining, (iv) Optimal feature selection with Mayfly algorithm, (v) Serial feature concatenation, and (vi) Binary classification and validation. This work considered the test image with dimension $896 \times 768 \times 3$ pixels. During the investigation, every picture is cropped into 25 slices and resized to $224 \times 224 \times 3$ pixels. This work implements the following stages; (i) BC detection with deep-features and (ii) BC recognition with concatenated features. In both the cases, a 5-fold cross validation is employed and the experimental investigation of this research confirms that the proposed work helped to achieve an accuracy of 91.39% with deep-feature and 95.56% with concatenation features.

Keywords: Breast cancer · Histology slide · ResNet18 · DWT features · Classification

1 Introduction

Cancer is one of the harsh illnesses in human; in which the cell in a particular organ/tissue will grow hysterically and spread to other body parts through blood stream. Even though

R. Chbeir et al. (Eds.): MIKE 2021, LNAI 13119, pp. 57–66, 2022.
https://doi.org/10.1007/978-3-031-21517-9_6

a considerable number of modern treatment facilities are accessible; the mortality rate due to cancer is gradually rising globally and the World Health Organization (WHO) report that, 9.6 million deaths are caused in 2018 globally because of the cancer [1].

In women community, the Breast Cancer (BC) is one of the harsh diseases and chief reason of death worldwide. Year 2020 report of WHO confirms that 2.3 million women are confirmed with BC and BC caused death is raised to 685000 globally. This report also stated that; before the end of year 2020, almost 7.8 million women living globally is diagnosed with BC in the past 5 years. This statement confirms the impact of the BC and its prevalent nature in women community.

The clinical level diagnosis of BC is commonly performed using medical imaging techniques, such as Ultrasound, Mammogram, MRI and thermal images. After confirming the presence of the BC in women, its stage is then examined using an invasive procedure called the needle biopsy (NB). During the NB examination, the necessary tissue section is extracted and then necessary clinical processing and examination process is implemented to confirm the stage (Benign/Malignant) of BC. During this procedure, histology slices are created using samples collected during NB and the slides are examined using a chosen procedure to confirm the stage of BC [2, 3].

In the literature, a number of Breast Histology Slide (BHS) examination procedures are discussed by the researches and most of these approaches employed the Deep-Leering (DL) techniques to get better disease detection. In these approaches, the whole BHS will be divided into small image groups (patches) and every patch is examined separately to identify the class.

The proposed research aims to implement a DL technique to detect the Benign/Malignant class BC from the BHS. In this work, initially, the performance of pre-trained DL schemes are inveterate using SoftMax classifier and the DL scheme which offers the best classification result for the BHS is chosen. Later, the necessary handcrafted-feature (HF) is extracted from the BHS using Discrete Wavelet Transform (DWT) and then a feature reduction process is implemented with the Mayfly-Optimization (MA) algorithm. The MA selected deep-feature (DF) and HF are then combined (DF+HF) to form a new feature sub-set; which is then considered to train and validate the binary classifiers. Based on the attained result from the classifier, the performance of the proposed BC detection system is confirmed.

In this research work, 1200 numbers (600 benign + 600 malignant) of resized BHS are considered for the assessment and the performance of the classifiers is confirmed using a 5-fold cross validation. The experimental result of this study confirms that the proposed work helped to get an accuracy of 91.39% with DF and 95.56% with DF+HF.

The chief contribution of this research includes:

- Comparison of the performance of pre-trained DL schemes on BHS image
- MA based DF and HF optimization
- Serial feature concatenation (DF+HF) and validation

Other sections of this work is prepared as follows; Sect. 2 presents earlier research work, Sect. 3 demonstrates the methodology, Sect. 4 and 5 presents experimental outcome and the conclusion respectively.

2 Related Research

Biomedical image supported automatic recognition of BC is commonly discussed by the researchers and the earlier work confirms the implementation of traditional and DL scheme based detection of BC with better accurateness. The BHS image based detection of the BC is also play a vital diagnostic role in clinics and the summary of few chosen BC investigative method with BHS is presented in Table 1.

Table 1. Summary of breast cancer evaluation with histology slides

Reference	Implemented assessment technique
Rawat et al. [4]	Implementation of DL scheme based assessment of BC from ER/PR/Her2 status from H&E images is presented
Rakha et al. [5]	Invasive BC examination based on visual histological assessment of morphological features is presented in this work
Egnell et al. [6]	Clinical grade assessment of breast histology specimen of Benign/Malignant breast lesions are presented and discussed
Kim et al. [7]	Employment of DL scheme based assessment of BC using frozen tissue section digital slide is discussed in this work
Duggento et al. [8]	Deep convolutional technique based assessment of digital pathology image is discussed
Wahab et al. [9]	BC detection from whole-slide histopathology picture estimation with multifaceted fused-CNN is presented and the attained results are discussed
Alzubaidi et al. [10]	Hybrid deep CNN based transfer learning approach is implemented to support optimizing the result of BC classification
Bianconi et al. [11]	Implementation of color-deconvolution and color-normalization based automatic categorization of histology pictures marked with hematoxylin/eosin is demonstrated
Ibrahim et al. [12]	Implementation of various Artificial Intelligence (AI) technique based examination of digital breast pathology is presented with appropriate examples

Table 1 outlines the research work proposed for detecting the BC using the histology images and this work also specified the need of the AI technique to detect the disease from the whole BHS or the patches obtained from the BHS. In this research, the whole BHS image is divided into 25 patches and resized to an image of dimension $224 \times 224 \times 3$ pixels and the created histology image dataset is examined using the pre-trained DL schemes, such as VGG16, VGG19, ResNet18, ResNet50 and ResNet101. Based on the classification accuracy, a DL scheme (ResNet18) is chosen and its features are then optimized with MA. The DWT features are also extracted from these images and optimal features are selected with MA. Finally a serial concatenation is then employed to combine the DF and HF (DF+HF) and the obtained feature vector is considered to train and validate the binary classifier.

3 Methodology

This division of paper presents the methodology employed in this research work and the different stages existing in this pipeline is depicted in Fig. 1.

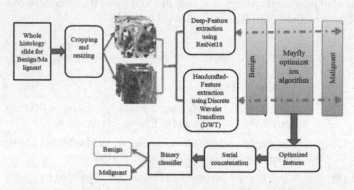

Fig. 1. Different stages of proposed breast cancer detection scheme

Figure 1 depicts the various phases available in the proposed pipeline and the collected Whole Histology Slide (WHS) is divided into 25 patches and resized to a chosen dimension. The features from these images are separately extracted with DL and DWT approach and then MA is employed to select the optimal features. The selected features are then concatenated serially to get a new one dimensional (1D) feature vector (DF+HF) and these features are then considered to train and validate the classifiers using a 5-fold cross validation.

3.1 Image Database

The collection of the appropriate medical image to validate the performance of the disease detection system is essential and in the proposed work, the necessary BHS are collected from [13]. This dataset consist 32 Benign and 26 malignant BH S with dimension of $896 \times 768 \times 3$ pixels and every image is then divided into 25 equal patches and all these patches are then resized into $227 \times 227 \times 3$ pixels. The number of images considered in this research is depicted in Table 2 and the sample test images of this database are shown in Fig. 2 and Fig. 3. Figure 2 depicts the sample Benign/Malignant image with possible patch traces and Fig. 3 presents the sample test images adopted in this work.

3.2 Feature Extraction

In this research, the necessary features from the test images are extracted using the DL scheme and the DWT approach. Initially, the ore-trained DL method is implemented to extract the necessary features from the test images. For each DL scheme, the following initial parameters are assigned; Adam optimizer, learning rate $= 1e-4$, dropout rate $= 50\%$, number of iterations $= 2000$ and iterations per epoch $= 40$. After the possible

Table 2. Information about the images considered in this research work

Class	Dimension	Images		
		Total	Training	Validation
Benign	$227 \times 227 \times 3$	600	420	180
Malignant	$227 \times 227 \times 3$	600	420	180

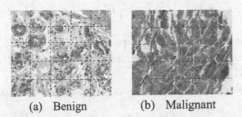

(a) Benign   (b) Malignant

Fig. 2. Whole slide dataset along with tracing for the patch

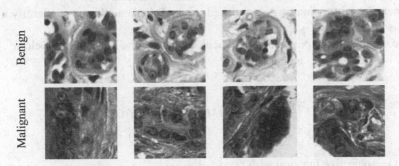

Fig. 3. Sample resized image considered for examination

dropout, the number of futures achieved will be $\approx 1 \times 1 \times 1024$ and these features are further reduced using the MA algorithm based optimization. The reduced features are then considered to train and validate the system with binary classifiers.

After getting the DL feature, the HF are then extracted using DWT practice and the necessary DWT features are then extracted from the images of coefficients, such as approximation, horizontal, vertical and diagonal. After extracting the required features, the optimal DWT features are obtained with the MA. These HF are then serially combined with DF to get concatenated features.

3.3 Feature Selection

Optimal feature selection from DF and HF are achieved with MA and the theoretical background and its implementation can be found in [14–17].

The parameters in MA consist identical male (A) and female (B) flies arbitrarily initialized in a D-dimensional exploration space and each fly is represented as follows; i

$= 1, 2, \ldots, N$. When MA run begins, each fly is randomly positioned in the elected space and allowed to unite towards the finest spot (G_{best}). The male is permissible to unite to G_{best} by varying its place and velocity. The meeting of a fly towards best outcome will be watched by the Cartesian-Distance (CD) with raise in iteration.

The mathematical expression for this process is depicted in Eqs. (1) to (5);

$$E_i^{t+1} = E_i^t + F_i^{t+1} \tag{1}$$

$$F_{i,j}^{t+1} = F_{i,j}^t + C_1 * e^{-\beta D_p^2}(P_{best_{i,j}} - E_{i,j}^t) + C_2 * e^{-\beta D_g^2}(G_{best_{i,j}} - E_{i,j}^t) \tag{2}$$

where E_i^t and E_i^{t+1} are primary and final positions, F_i^{t+1} and $F_{i,j}^{t+1}$ primary and final velocities correspondingly, The other parameters are assigned as follows; $C_1 = 1, C_2 = 1.5, \beta = 2$ and D_p and D_g are CD.

The velocity revising throughout this procedure can be defined as;

$$F_{i,j}^{t+1} = F_{i,j}^t + d * R \tag{3}$$

where nuptial dance value (d) $= 5$ and R $=$ random numeral $[-1, 1]$.

When the exploration of A is finished, every B is then allowed to identify a male which is converged at G_{best}.

The expression for position and velocity update for the F is depicted below.

$$E_i'^{t+1} = E_i'^t + F_i'^{t+1} \tag{4}$$

$$F_{i,j}'^{t+1} = \begin{cases} F_{i,j}'^t + C_2 e^{-\beta D_{mf}^2}(M_{i,j}^t - Y_{i,j}^t) & \text{if } O(F_i) > O(M_i) \\ F_{i,j}'^t + W * r & \text{if } O(F_i) \le O(M_i) \end{cases} \tag{5}$$

where O $=$ objective function with maximal value.

Other necessary information on the MA can be found in [18–20].

In this work, the MA is considered to find the optimal features in DF and HF of Benign/Malignant class images are optimized based on the CD (the features with maximal CD is selected as finest features).

3.4 Classification and Validation

The performance of the disease recognition structure depends mainly on the performance measures achieved and in this work. Further, the choice of the classifier also plays a major role in getting better measures. In this work, the binary classification is implemented with SoftMax, Naïve-Bayes (NB), Decision-Tree (DT), Random-Forest (RF), k-Nearest Neighbors (KNN), and Support-Vector-Machine (SVM) with linear kernel. In this work, a 5-fold cross validation is employed and the necessary measures, such as True-Positive (TP), False-Negative (FN), True-Negative (TN) and False-Positive (FP) are initially computed and from these vales, accuracy (AC), precision (PR), sensitivity (SE), specificity (SP) and negative predictive value (NPV) are computed [21, 22].

4   Result and Discussion

This section of the paper presents the experimental result achieved using the workstation This part of work demonstrates the results achieved on a workstation; Intel i7 2.9 GHz processor with 20 GB RAM and 4 GB VRAM equipped with Matlab®.

Initially, the pre-trained DL scheme is employed to extract the DF using the methods such as VGG16, VGG19, ResNet18, ResNet50 and ResNet101and the achieved result is compared and validated to find the suitable DL scheme for the chosen database.

The considered DL scheme is trained using the test images and after achieving the necessary training, the performance is tested and validated with other considered images.

Figure 4 presents the intermediate layer output achieved with ResNet18. Figure 4(a) and (b) presents the results of convolutional layer and Fig. 4(c) presents the MaxPool layer result. Table 3 depicts the outcome of ResNet19 for SoftMax classifier with a 5-fold cross validation. Similar procedure is the employed for other DL schemes and the attained result is depicted in Table 4. This table confirms the merit of ResNEt18 on the chosen image database. Its performance is also graphically depicted in Fig. 5. Figure 5(a) shows the classifier accuracy for various folds and Fig. 5(b) presents the Glyph-plot to demonstrate the overall performance of DL schemes. This image also verified the eminence of resNet18 over other schemes.

The handcrafted features are then extracted from the DWT treated images and the extracted features are then optimized with the procedure discussed as in subsect. 3.3. Optimized DWT feature is then combined with deep-features of ResNet18 and then the classification task is repeated and the attained results are presented in Table 5 (Fig. 6).

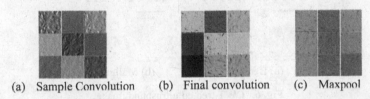

(a)   Sample Convolution   (b)   Final convolution   (c)   Maxpool

Fig. 4. Sample results achieved from various layers of ResNet18

Table 3. Results achieved during 5-fold cross validation

Classifier	TP	FN	TN	FP	AC	PR	SE	SP	NPV
Trial1	156	24	155	25	86.39	86.19	86.67	86.11	86.59
Trial2	157	23	160	20	88.06	88.70	87.22	88.89	87.43
Trial3	159	21	159	21	88.33	88.33	88.33	88.33	88.33
Trial4	166	14	163	17	91.39	90.71	92.22	90.56	92.09
Trial5	164	16	162	18	90.56	90.11	91.11	90.00	91.01

Table 5 confirms that the classification result achieved with DT classifier for the concatinated features and the sample results acieved during this experiment is presented

Table 4. Classification results achieved with pre-trained DL schemes

Classifier	Selected features	TP	FN	TN	FP	AC	PR	SE	SP	NPV
VGG16	583	162	18	161	19	89.72	89.50	90.00	89.44	89.94
VGG19	614	159	21	162	18	89.17	89.83	88.33	90.00	88.52
ResNet18	517	166	14	163	17	91.39	90.71	92.22	90.56	92.09
ResNet50	448	161	19	164	16	90.28	90.96	89.44	91.11	89.62
ResNet101	504	162	18	161	19	89.72	89.50	90.00	89.44	89.94

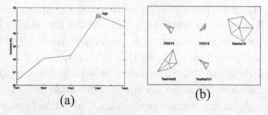

(a)   (b)

Fig. 5. Classification accuracy and the performance achieved with pretrained DL scheme

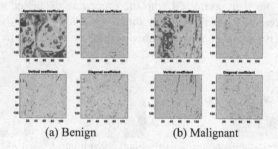

(a) Benign   (b) Malignant

Fig. 6. DWT treated ultrasound image

Table 5. Result achieved with ResNet18 for different classifiers

Classifier	TP	FN	TN	FP	AC	PR	SE	SP	NPV
SoftMax	169	11	170	10	94.17	94.41	93.89	94.44	93.92
NB	171	9	169	11	94.44	93.96	95.00	93.89	94.94
DT	171	9	173	7	95.56	96.07	95.00	96.11	95.05
RF	170	10	172	8	95.00	95.51	94.44	95.56	94.51
KNN	172	8	169	11	94.72	93.98	95.56	93.89	95.48
SVM	170	10	171	9	94.72	94.97	94.44	95.00	94.48

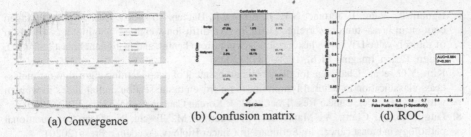

      (a) Convergence   (b) Confusion matrix   (d) ROC

Fig. 7. Final result achieved with ResNet18 for DT classifier

in Fig. 7. Figure 7(a) shows the convergence and Fig. 7(b) and (c) presents confusion matrix and Receiver Operating Characteristic (ROC) curve achieved for this classifier. From this study, it is confirmed thet the ResNet18 scheme with concatinated feature helped to achieve a better result on the considered dataset and helped to achieve a better accuracy during the classification od benign/malignant class.

5 Conclusion

Medical image supported disease detection is a necessary procedure in most of the hospitals to detect the disease in its early phase. To improve the detection accuracy, DL schemes are widely employed in the literature and in these work pre-trained DL methods are employed to classify the breast histology images into benign/malignant class. The proposed work considered the deep-feature and concatenated feature based disease detection. In this work, feature optimization is implemented with MA and the optimized features are considered to detect the disease class with BIS. In this work the DWT feature is considered as the chosen handcrafted-feature and concatenated feature helps to get a better result compared to Deep-features. In this work, ResNet18 classifier is considered with a binary classifier and a 5-fold cross validation is also executed to get a better result. In this work, DT classifier executed with concatenated feature is helped to get a better disease detection (accuracy >95.56%) compared to other methods. This result confirms that the proposed scheme can be used to evaluate the clinical grade histology images in future.

References

1. https://www.who.int/news-room/fact-sheets/detail/breast-cancer
2. Raja, N., Rajinikanth, V., Fernandes, S.L., Satapathy, S.C.: Segmentation of breast thermal images using Kapur's entropy and hidden Markov random field. J. Med. Imaging Health Inform. **7**(8), 1825–1829 (2017)
3. Fernandes, S.L., Rajinikanth, V., Kadry, S.: A hybrid framework to evaluate breast abnormality using infrared thermal images. IEEE Consum. Electron. Mag. **8**(5), 31–36 (2019)
4. Rawat, R.R., et al.: Deep learned tissue "fingerprints" classify breast cancers by ER/PR/Her2 status from H&E images. Sci. Rep. **10**(1), 1–13 (2020)
5. Rakha, E.A., et al.: Visual histological assessment of morphological features reflects the underlying molecular profile in invasive breast cancer: a morphomolecular study. Histopathology **77**(4), 631–645 (2020)

6. Egnell, L., Vidić, I., Jerome, N.P., Bofin, A.M., Bathen, T.F., Goa, P.E.: Stromal collagen content in breast tumors correlates with in vivo diffusion-weighted imaging: a comparison of multi b-value DWI with histologic specimen from benign and malignant breast lesions. J. Magn. Reson. Imaging **51**(6), 1868–1878 (2020)

7. Kim, Y.G., et al.: Challenge for diagnostic assessment of deep learning algorithm for metastases classification in sentinel lymph nodes on frozen tissue section digital slides in women with breast cancer. Cancer Res. Treat.: Off. J. Korean Cancer Assoc. **52**(4), 1103 (2020)

8. Duggento, A., Conti, A., Mauriello, A., Guerrisi, M., Toschi, N.: Deep computational pathology in breast cancer. In: Seminars in Cancer Biology. Academic Press (2020)

9. Wahab, N., Khan, A.: Multifaceted fused-CNN based scoring of breast cancer whole-slide histopathology images. Appl. Soft Comput. **97**, 106808 (2020)

10. Alzubaidi, L., Al-Shamma, O., Fadhel, M.A., Farhan, L., Zhang, J., Duan, Y.: Optimizing the performance of breast cancer classification by employing the same domain transfer learning from hybrid deep convolutional neural network model. Electronics **9**(3), 445 (2020)

11. Bianconi, F., Kather, J.N., Reyes-Aldasoro, C.C.: Experimental assessment of color deconvolution and color normalization for automated classification of histology images stained with hematoxylin and eosin. Cancers **12**(11), 3337 (2020)

12. Ibrahim, A., et al.: Artificial intelligence in digital breast pathology: techniques and applications. Breast **49**, 267–273 (2020)

13. https://academictorrents.com/details/b79869ca12787166de88311ca1f28e3ebec12dec

14. Khan, M.A., et al.: Computer-aided gastrointestinal diseases analysis from wireless capsule endoscopy: a framework of best features selection. IEEE Access **8**, 132850–132859 (2020)

15. Wang, Y., et al.: Classification of mice hepatic granuloma microscopic images based on a deep convolutional neural network. Appl. Soft Comput. **74**, 40–50 (2019)

16. Rajinikanth, V., Kadry, S., Taniar, D., Damaševičius, R., Rauf, H.T.: Breast-cancer detection using thermal images with marine-predators-algorithm selected features. In: 2021 Seventh International conference on Bio Signals, Images, and Instrumentation (ICBSII), pp. 1–6. IEEE (2021)

17. Cheong, K.H., et al.: An automated skin melanoma detection system with melanoma-index based on entropy features. Biocybern. Biomed. Eng. **41**(3), 997–1012 (2021)

18. Zervoudakis, K., Tsafarakis, S.: A mayfly optimization algorithm. Comput. Ind. Eng. **145**, 106559 (2020)

19. Kadry, S., Rajinikanth, V., Koo, J., Kang, B.G.: Image multi-level-thresholding with Mayfly optimization. Int. J. Electr. Comput. Eng. (2088–8708) **11**(6), 5420–5429 (2021)

20. Gao, Z.M., Zhao, J., Li, S.R., Hu, Y.R.: The improved mayfly optimization algorithm. In: Journal of Physics: Conference Series, vol. 1684, no. 1, p. 012077. IOP Publishing (2020)

21. Ahuja, S., Panigrahi, B.K., Dey, N., Rajinikanth, V., Gandhi, T.K.: Deep transfer learning-based automated detection of COVID-19 from lung CT scan slices. Appl. Intell. **51**(1), 571–585 (2020). https://doi.org/10.1007/s10489-020-01826-w

22. Rajinikanth, V., Joseph Raj, A.N., Thanaraj, K.P., Naik, G.R.: A customized VGG19 network with concatenation of deep and handcrafted features for brain tumor detection. Appl. Sci. **10**(10), 3429 (2020)

Recent Trends in Human Re-identification Techniques – A Comparative Study

J. Stella Janci Rani$^{(\boxtimes)}$ and M. Gethsiyal Augasta

Research Department of Computer Science, Kamaraj College (affiliated to Manonmaniam Sundaranar University, Tirunelveli), Thoothukudi, Tamilnadu, India
`jstellajara17@gmail.com`

Abstract. Human re-identification is an essential one in recent days for automated video crime avoidance. Human re-identification is collecting impetus owing to its purpose in security and crime inquiry. Handcrafted features namely, size, color and texture histograms with small-scale assessment were used for re-identification in early times. But nowadays many deep learning-based methods have emerged to provide excellent performance in a very wide range of applications together with image annotation, face recognition and speech recognition. Deep learning features are heavily dependent on large-scale labeling of samples. This paper concentrates on analyzing different existing human re-identification using handcrafted features, deep features and combined features. The main objective of this study is to provide a detailed insight to the reader with the different aspects of re-identification methodologies and future direction of researches. The experimental evaluations on recent re-identification methods have been presented for a better understanding of the performance of existing human re-identification methods on different benchmark datasets namely VIPeR, Market1501, and CUHK03.

Keywords: Human/Person Re-identification · Handcrafted features · Deep features · Distance metric learning · Feature extraction · Deep learning

1 Introduction

One of the foremost significant facts of the intelligent surveillance system, which has been reviewed in the literature, is Human re-identification. Human re-identification technology is an important form of biometrics for recognition. It can recognize an individual based on their overall appearance, even when their face is not visible. In general, it is the process of recognizing individuals over various camera views in different locations under the condition of large illumination variations. Human re-identification plays a major role in criminal investigation. Human re-identification techniques are demanding due to the dearth of spatial and temporal constraints and large visual appearance. These constraints are caused by variations in body pose, view angle, lighting, and background clutter, scenarios across time, occlusion, and cameras [1]. In other words, appearance-based human re-identification depends on the information provided by the visual appearance of the human body and the clothing. It is an extremely difficult problem since human

R. Chbeir et al. (Eds.): MIKE 2021, LNAI 13119, pp. 67–77, 2022.
https://doi.org/10.1007/978-3-031-21517-9_7

appearance usually exhibits giant variations across completely different cameras. The purpose of human re-identification is to identify the correct match from a large collection of gallery images describing the same person captured by disjoint camera views.

The human re-identification technique has three main requirements. First, there is a requirement to conclude which parts should be segmented and compared. Second, there is a requirement to form invariant signatures for comparing the corresponding divisions. Third, an acceptable metric should be applied to discriminate the signatures. Under the premise that the looks of the human remain unchanged, most re-identification approaches are intended. Based on this supposition, local descriptors like color and texture are well thought-about to use the sturdy signatures of images. Researchers have proposed various techniques for human re-identification. However, every technique has its own restrictions. For instance, human re-identification in appearance-based methods must deal with several challenges such as variations within an illumination situation, poses and occlusions across time and cameras. In this study, various human re-identification techniques have been analysed and experimentally evaluated with two outdoor environment datasets namely VIPeR, Market 1501 and indoor environment dataset CUHK03.

2   Methodologies for Human Re-identification

Feature representation and distance metrics are two elementary components necessary for human re-identification systems [1]. Here, feature representation being significantly necessary as a result of its styles of metric learning. The aim of feature representation is to develop the discrimination and the robust appearance of the identical pedestrian across diverse camera views.

2.1   Feature Representation

The representation of the feature is an essential one because it forms the basis for metric learning. In conjunction with the image processing technique, feature extraction commences with a collection of measured data and then creates a series of derivative values that are intended to be informative and non-redundant. Most of the feature extraction schemes are usually utilized in the re-identification fiction method. The extracted features are generally divided into two parts, namely handcrafted features and deep learned features.

Handcrafted Features: Handcrafted features like color, light, texture and a human's appearance are commonly used for human re-identification [2]. Human images are low in resolution and have large variations in the pose. Thus, it has been shown that colour details such as colour histograms and colour name descriptors are the most significant cue for individual re-identification. Since numerous individuals of similar colour cannot be adequately separated, textural descriptors such as Local Binary Pattern (LBP) and the responses of filter deposits are also combined with color descriptors. Human re-identification methods [3–6] include color histograms [7], LBP [6] and Scale-Invariant Feature Transform (SIFT) [3]. In ELF [7], color histograms within the RGB, YCbCr, and HS color spaces, and texture filters are used [8, 9]. In SIFT [3] the Illumination changes, viewpoint, pose variations and scene occlusion are the most important for re-identification challenges. In HistLBP [10], color histograms within the RGB, YCbCr, and HS color spaces and texture histograms from local binary patterns (LBP) [6] features are computed. In DenseColorSIFT [3], every image is densely divided into patches, and color histograms and SIFT features are extracted from each patch, Dense SIFT handles viewpoint and illumination amendment, SIFT descriptor is employed as a complementary feature to color histograms. Spatial co-occurrence representation [11] and mixtures of multiple features [12] have been used.

Liao et al. [13] have proposed Local Maximal Occurrence [LOMO] to represent every pedestrian image as a high dimensional feature. LOMO feature extraction is a method, which is shown to be robust against viewpoint changes and illumination variation. HSV color histograms and scale-invariant LBP [14] features are extracted from an image created by a multi-scale Retinex algorithm [15], and maximally pooled along the identical horizontal strip. As an example, Kostinger et al. [16] have divided images into regular grids from which color and texture features were extracted from overlapping blocks. By exploiting regular and irregular perceptual principles, Farenzena et al. [17] extracted three feature varieties to model distant aspects of human appearance as well as Maximally Stable Color Regions (MSCR), weighted color histograms, and Recurrent High Structured Patches (RHSP). These strategies were strong to low resolution and occluded images, pose, illumination, and viewpoint changes. Beneath totally different circumstances, Liu et al. [18] have projected different sets of features to determine significant features for person re-identification. They have used an unsupervised approach to weigh the importance of various features. However, traditional color information may not be the foremost effective way to describe the color. Hence, Kuo et al. [19] have applied semantic color names to explain pedestrian images and calculate chance distributions on these basic colors as image descriptors. The improved LOMO [20] is merged with the midlevel semantic color names to deal with the viewpoint changes for human re-identification downside.

Deep Learned Features: In addition to handcrafted features, efforts have also been applied to mechanically learn features for human re-identification. Gray et al. [7] have projected the employment of AdaBoost to find out strong representations from a collection of localized features. It has been reported that deep learning is "state-of-the-art" for a wide variety of tasks as well as image annotation, face recognition and speech recognition. Zhou et al. [21] have learned deep features for scene recognition and found state-of-the-art results on scene-centric datasets. The accomplishment of Convolution Neural Network (CNN) [22] is to automatically learn the deep features and has lately received attention. Yi et al. [23, 24] have used a "Siamese" deep neural network to increase human re-identification performance the projected model collectively learnt the color feature, texture feature and metrics in an exceedingly unified framework.

Li et al. [25] have proposed a deep learning framework FPNN to educate the filter pairs that tried to automatically encode the photometric transforms across cameras, such that geometric and photometric transforms, misalignment, occlusions, and background clutter were with efficiency handled. Meanwhile, Shi et al. [26] have projected a unique deep architecture to learn the Mahalanobis metric with a weight constraint to achieve good generalizability of the learned metric. Ahmed et al. [27] have presented a deep architecture for human re-identification with original components. This architecture has a layer for computing cross-input neighborhood variations to capture relationships between mid-level features calculated separately from the two input human images. EDF [28] is a Triplet-based model to learn the deep features. Harmonious Attention Convolutional Neural Network HA-CNN [29] performs joint learning of human re-identification concentration selection and feature representations in one after the other fashion. Virtual CNN Branching [30] allows a network to demonstrate ensemble behaviour without additional parameters. However, these automatically learned features are reliant on large-scale labeling of samples.

Combined Handcrafted and Deep Features: By analyzing all recent works, in order to learn the deep function, Tao et al. [28] implemented a triplet-based EDF model. Then, it extracted the LOMO handcrafted feature. Next, to approximate the two distances, XQDA metric models were trained on the EDF and the LOMO, respectively. After that, to re-identify the person, the final distance is computed from the above two distances. SLFDLF [31] method is the combination of Splitted LOMO feature and deep learned feature. Normally LOMO features are extracted from the whole image; instead of taking global features, the local features have more discriminative nature to describe the human structural information. In this method, initially, the image is horizontally partitioned into grids based upon the size and then the LOMO features are extracted from each grid. Then the extracted LOMO features are combined with deep features to represent that image and XQDA metric learning model is used to evaluate the combination of Splitted LOMO features and deep learnt features (SLFDLF). Jayapriya et al., [32] have introduced a method where SILTP is applied to an image's RGB channel, which is identified for its invariant texture description of illumination.

2.2 Distance Metric Learning Methodologies

Distance Metric Learning Methodologies define a distance between every try of components of a collection. Metric learning intends to develop a discriminant matching model to gauge sample similarity [33–39]. It has received widespread interest for human re-identification. It is reflected on 18 metric learning methods [1] that are usually used by the re-identification community. Fisher-type optimization which includes Fisher Discriminant Analysis (FDA) [40], Local Fisher Discriminant Analysis (LFDA) [41], Marginal Fisher Analysis (MFA) [42], Cross-view Quadratic Discriminant Analysis (XQDA) [13], and discriminative Null Space Learning (NFST) [43] looks for minimizing the within-class data disperse while maximizing between-class data disperse. Davis et al. [44] have proposed Information-Theoretic Metric Learning (ITML) foundation on the Mahalanobis distance and aggravated by minimizing the degree of difference relative entropy between two multivariate Gaussian distributions beneath metric function restriction.

Zhao et al. [45] have exploited a pairwise saliency distribution relationship between images and proposed a saliency matching strategy called salMatch, for human re-identification. Weinberger et al. [46] have proposed the Large Margin Nearest neighbour

(LMNN) learning method which set up a perimeter for the target neighbours and punished those invading the perimeter. Zheng et al. have proposed Probabilistic Relative Distance Comparison (PRDC) [47] to maximize the probability of a pair of the accurate match having a smaller distance than that of an incorrect match pair. Keep-It-Simple-and-Straightforward Metric (KISSME) [16] and Pairwise Constrained Component Analysis (PCCA) [48] learn Mahalanobis-type distance functions using the deviation of the essential pairwise constraints principle. By adopting the statistical perspective, Kostinger et al. [16] have proposed a simple yet robust method called KISS to learn the metrics. The metrics are KISSME [16], kPCCA [48], kLFDA [10], rPCCA [10], and kMFA [10] kernelize PCCA, LFDA, and MFA, respectively. To map the kernelized features into a common subspace, KCCA [49] adopts canonical correlation analysis. RankSVM [50] seeks out the best weights for the similarity measurements. SVMML [51] learns locally-adaptive decision functions that are learned in a very large-margin SVM framework.

3   Performance Analysis

In this section, the results of some existing human re-identification methods such as PCT-CNN [32], DMVFL [28], DECAMEL [57], CDML [26], LOMO + XQDA [13], Improved DML [23], EDF [28], UTAL [56], k-reciprocal Encoding [58], DSML [62], PersonNet [60] and DML [24] are compared with each other on well-known three real datasets and their performances are evaluated. The detailed description of the datasets is shown in Table 1.

Table 1. Properties of three real datasets

Properties	Datasets		
	VIPeR	Market1501	CUHK03
#of classes	632	1501	1467
#of images	1264	32,668	14,097
#of the training set	316	750	1367
#of the testing set	316	751	100
#of image sizes	128×48	128×64	Vary

3.1   Comparative Analysis

The rank obtained on testing data with the most popular re-identification methods in our study has been compared to evaluate their performance. A search user is matched against a set of gallery humans for producing a ranked list according to their matching similarity, characteristically assuming the correct match is assigned to the top rank preferably Rank1.

Comparison on VIPeR: VIPeR [7] dataset is compared with other methods such as PCT-CNN [32], DMVFL [28], CDML [26], LOMO + XQDA [13], Improved DML [23], EDF [28], DECAMEL [57] and DML [24]. Table 2 shows the comparison results of the VIPeR dataset. From Table 2, it can be observed that the methods PCT-CNN [32] and DMVFL [28] which are the combination of handcrafted feature and deep learned feature outperforms. The Constrained Deep Metric Learning (CDML) is the CNN-based method to learn a discriminative metric with well-thought-of robustness to the over-fitting problem in person re-identification. The LOcal Maximal Occurrence (LOMO) handcrafted features are frequently used in person re-identification and they are considered as a high-dimensional function for feature extraction technique from each pedestrian image [13]. It extracts Scale Invariant Local Ternary Pattern (SILTP) and HSV color histograms to form high level descriptors. It also examines the horizontal occurrence of local geometric traits and maximizes their occurrences when perspectives vary. The DML method can learn the colour feature, texture feature, and metric in a single framework by employing a "Siamese" deep neural network.

Table 2. Comparison results of the VIPeR dataset

Method	Rank 1	Year
PCT-CNN [32]	47.15	2020
DMVFL [28]	46.4	2017
EDF [28]	31.2	2017
CDML [26]	40.91	2015
DECAMEL [57]	34.15	2019
LOMO + XQDA [13]	40	2015
Improved LOMO + SCNCD [20]	42.72	2015
Improved DML [23]	34.4	2014
DML [24]	28.3	2014

Comparison on Market1501: Market1501 [52] dataset is compared with re-identification methods such as SLFDLF [31] DLF [31], LFDLF [31] DLPR [54], k-reciprocal Encoding [58], BoW [52], eSDC [61], SDALF [59] and CNN based methods like DSML [62], MAR [55], UTAL [56], PersonNet [60] and DECAMEL [57]. The comparison results of the Market1501 dataset are listed in Table 3. Results in Table 3 depict that most of the deep learning or deep feature based methods outperform the other methods. The deep feature based SLFDLF method obtained the top matching retrieval rate 85.35% compared to other methods.

Table 3. Comparison results of Market1501 dataset

Method	Rank 1	Year
SLFDLF [31]	85.35	2020
SLF [LOMO + XQDA] [31]	82.95	2020
LFDLF [31]	84.15	2020
LF [LOMO + XQDA] [31]	81.92	2020
DLF [31]	64.61	2020
UTAL [56]	56.3	2020
DSML [62]	84.4	2019
MAR [55]	67.7	2019
DECAMEL [57]	60.24	2019
DLPR [54]	81	2017
k-reciprocal Encoding [58]	77.11	2017
PersonNet [60]	37.21	2016
BoW [52]	34.4	2015
eSDC [61]	33.5	2013
SDALF [59]	20.5	2013

Comparison on CUHK03: CUHK03 [53] dataset are analyzed with the handcrafted feature and CNN based feature methods such as LOMO + XQDA [13], DECAMEL [57], VCFL [64], k-reciprocal Encoding [58], Deep Part-Aligned [54], CSBT [63], SDALF [59], FPNN [53], OSNet [65] DSML [62], UTAL [56], PersonNet [60]. Table 4 shows the comparison results on CUHK03 dataset. Here deep learning-based method OSNet [65] performs outstandingly than other methods.

Table 4. Comparison results of CUHK03 dataset

Method	Rank 1	Year
OSNet [65]	72.3	2021
UTAL [56]	69.2	2020
DECAMEL [57]	45.82	2019
DSML [62]	88	2019
VCFL [64]	70.36	2019
k-reciprocal Encoding [58]	69.9	2017

(continued)

Table 4. (*continued*)

Method	Rank 1	Year
Deep Part-Aligned [54]	85.4	2017
CSBT [63]	55.5	2017
PersonNet [60]	64.8	2016
LOMO + XQDA [13]	52.20	2015
FPNN [53]	20.65	2017
SDALF [59]	5.6	2013

4   Conclusion

In this paper, a survey of algorithms for human re-identification has been specified. Most of the previous works have frequently used handcrafted features such as color, texture histogram and human appearance. The majority of handcrafted features have the drawback of not being immediately applicable to practical issues. In addition to handcrafted features, recently there have been attempts to learn features automatically for human re-identification by deep learning. Moreover, in some researches, deep and handcrafted features are flawlessly integrated to re-identify the human. From this present survey, it is observed that the human re-identification task has several challenges such as body misalignment, pose deformation, missing parts, different viewing angles, occlusion, etc. Even though many works are available for human re-identification, there still lies a gap in terms of proposing a new unified model that addresses all the listed challenges.

In addition to this survey, some of the recent re-identification methods based on handcrafted and deep features are evaluated on three real datasets such as VIPeR, Market1501, and CUHK03 with regard to their top matching results. The experimental results depict that the deep features or deep learning based methods will dominate in future for solving the human re-identification problems than the other machine learning methods.

References

1. Karanam, S., Gou, M., Wu, Z., Rates-Borras, A., Camps, O., Radke, R.J.: A systematic evaluation and benchmark for person re-identification: features, metrics, and datasets. IEEE Trans. Pattern Analy. Machine Intelli. **41**(3) (2018)
2. Lu, X., Zheng, X., Li, X.: Latent semantic minimal hashing for image retrieval. IEEE Trans. Image Proc. **26**(1), 355–368 (2017)
3. Lowe, D.G.: Distinctive image features from scale-invariant key points. Int. J. Comp. Vision **60**(2), 91–110 (2004)
4. Tan, S., Zheng, F., Liu, L., Han, J.: Dense invariant feature-based support vector ranking for cross-camera person re-identification. IEEE Trans. Circuits and Systems for Video Technol. (2016)
5. Yang, Y., Deng, C., Tao, D., Zhang, S., Liu, W., Gao, X.: Latent Max-Margin Multitask Learning with Skelets for 3-D Action Recognition. IEEE Transactions on Cybernetics **47**(2), 439–448 (2017)

6. Ojala, T., Pietikainen, M., Maenpa, T.: Multi resolution gray-scale and rotation invariant texture classification with local binary patterns. IEEE Trans. Pattern Analysis and Machine Intelli. **24**(7), 971–987 (2002)

7. Gray, D., Tao, H.: Viewpoint invariant pedestrian recognition with an ensemble of localized features. In: Proceedings of the European Conference on Computer Vision [ECCV], pp. 262–275 (2008)

8. Schmid, C.: Constructing models for content-based image retrieval. In: CVPR (2001)

9. Fogel, A., Sagi, D.: Gabor filters as texture discriminator. Biological cybernetics **61**(2), 103–113 (1989)

10. Xiong, F., et al.: Person re-identification using kernel-based metric learning methods. In: ECCV (2014)

11. Wang, X., Doretto, G., Sebastian, T., Rittscher, J., Tu, P.: Shape and appearance context modelling. In: Proceedings of the International Conference on Computer Vision, pp. 1–8 (2007)

12. Paisitkriangkrai, S., Shen, C., Anton, V.D.H.: Learning to rank in person re-identification with metric ensembles. In: Proceedings of the Conference on Computer Vision and Pattern Recognition, pp. 1846–1855 (2015)

13. Liao, S., Hu, Y., Zhu, X., Li, S.Z.: Person re-identification by local maximal occurrence representation and metric learning. In: Proceedings of the IEEE Conference on Computer Vision and Pattern Recognition, pp. 2197–2206 (2015)

14. Liao, S., Zhao, G., Kellokumpu, V., Pietikäinen, M., Li, S.Z.: Modeling pixel process with scale invariant local patterns for background subtraction in complex scenes. In: CVPR (2010)

15. Jobson, D.J., Rahman, Z.-U., Woodell, G.A.: A multi scale retinex for bridging the gap between color images and the human observation of scenes. T-IP **6**(7), 965–976 (1997)

16. Kostinger, M., Hirzer, M., Wohlhart, P., Roth, P.M., Bischof, H.: Large scale metric learning from equivalence constraints. In: Proceedings of the Conference on Computer Vision and Pattern Recognition, pp.2288–2295 (2012)

17. Farenzena, M., Bazzani, L., Perina, A., Murino, V., Cristani, M.: Person re-identification by symmetry-driven accumulation of local features. In: Proceedings of the IEEE Conference on Computer Vision and Pattern Recognition, pp. 2360–2367 (2010)

18. Liu, C., Gong, S., Loy, C.C., Lin, X.: Person re-identification: What features are important? In: Proceedings of the IEEE Conference on Computer Vision, pp. 391–401 (2012)

19. Kuo, C.H., Khamis, S., Shet, V.: Person re-identification using semantic color names and rank boost. In: Proceedings of the IEEE Winter Conference on Applications of Computer Vision, pp. 281–287 (2013)

20. Song, M., Liu, C., Ji, Y., Dong, H.: Person re-identification by improved local maximal occurrence with color names. In: 8th International Congress on Image and Signal Processing, pp. 675–679 (2015)

21. Zhou, B., Agata, L., Xiao, J., Antonio, T., Aude, O.: Learning deep features for scene recognition using places database. In: Proceedings of the Conference on Neural Information Processing Systems, pp. 487–495 (2015)

22. Opitz, M., Waltner, G., Poier, G., Possegger, H., Bischof, H.: Gridloss: detecting occluded faces. In: Proceedings of the European Conference on Computer Vision, pp. 386–402 (2016)

23. Yi, D., Lei, Z., Liao, S., Li, S.Z.: Deep metric learning for practical person re-identification. In: Proceedings of the International Conference on Pattern Recognition, pp.34–39 (2014)

24. Yi, D., Lei, Z., Liao, S., Li, S.Z.: Deep metric learning for person re identification. In: 22nd International Conference on Pattern Recognition, pp. 34–39 (2014)

25. Li, W., Zhao, R., Xiao, T., Wang, X.: Deepreid: deep filters pairing neural network for person re-identification. In: Proceedings of the Conference on Computer Vision and Pattern Recognition, pp. 152–159 (2014)

26. Shi, H., Zhu, X., Liao, S., Lei, Z., Yang, Y., Li, S.Z.: Constrained deep metric learning for person re-identification. In: Computer Vision and Pattern Recognition arXiv preprintar Xiv: 1511.07545 (2015)
27. Ahmed, E., Jones, M., Marks, T.K.: An improved deep learning architecture for person re-identification. In: Proceedings of the Conference on Computer Vision and Pattern Recognition, pp. 3908–3916 (2015)
28. Tao, D., Guo, Y., Yu, B., Pang, J., Yu, Z.: Deep multi-view feature learning for person re-identification. In: IEEE Transactions on Circuits and Systems for Video Technology, pp. 1–1 (2017)
29. Li1, W., Zhu, X., Gong, S.: Harmonious attention network for person re identification. In: Proceedings of the IEEE Conference on Computer Vision and Pattern Recognition (CVPR), pp. 2285–2294 (2018)
30. Gong, A., Qiu, Q., Sapiro, G.: Virtual CNN Branching: Efficient Feature Ensemble for Person Re-Identification. ArXiv: 1803.05872v1 (2018)
31. Stella, J., Rani, J., Gethsiyal Augasta, Dr.M.: A deep learning model for human re-identification with split LOMO and deep features using XQDA. Test Engineering & Management, pp. 13776–13786 (2020). ISSN: 0193-4120
32. Jayapriya, K., Jeena Jacob, I., Ani Brown Mary, N.: Person re-identification using prioritized chromatic texture (PCT) with deep learning. Multimedia Tools and Applications (2020)
33. Cong, Y., Fan, B., Liu, J., Luo, J., Yu, H.: Speeded-Up Low-Rank Online Metric Learning for Object Tracking. IEEE Transactions on Circuits and System for Video Technique 25(6), 922–934 (2015)
34. Johnson, D., Xiong, C., Corso, J.: Semi-supervised nonlinear distance metric learning via forests of max-margin cluster hierarchies. IEEE Trans. Knowl. Data Eng. 28(4), 1035–1046 (2016)
35. Li, X., Mou, L., Lu, X.: Surveillance video synopsis via scaling down objects. IEEE Trans. Image Proc. 25(2), pp. 740–755 (2016)
36. Bellet, A., Habrard, A., Sebban, M.: Metric Learning. Synthesis Lectures on Artificial Intelligence and Machine Learning 9(1), 1–151 (2015)
37. Lin, L., Lu, Y., Li, C., Cheng, H., Zuo, W.: Detection-free multiobject tracking by reconfigurable inference with bundle representations. IEEE Transactions on Cybernetics 46(11), 2447–2458 (2016)
38. Wattanachote, K., Shih, T.K.: Automatic dynamic texture transformation based on new motion coherence metric. IEEE Trans. Circu. Sys. Video Technol. 26(10), 1805–1820 (2016)
39. Xu, C., Tao, D., Xu, C.: Multi-view intact space learning. IEEE Trans. Pattern Analy. Machine Intelli. 37(12), 2531–2544 (2015)
40. Fisher, R.A.: The use of multiple measurements in taxonomic problems. Annals of eugenics (AE) 7(2), 179–188 (1936)
41. Pedagadi, S., Orwell, J., Velastin, S., Boghossian, B.: Local fisher discriminant analysis for pedestrian re-identification. In: CVPR (2013)
42. Yan, S., Xu, D., Zhang, B., Zhang, H.-J., Yang, Q., Lin, S.: Graph embedding and extensions: a general framework for dimensionality reduction. IEEE Transactions on Pattern Analysis and Machine Intelligence 29(1), 40–51 (2007)
43. Zhang, L., Xiang, T., Gong, S.: Learning a discriminative null space for person re-identification. In: CVPR, pp. 1239–1248 (2016)
44. Davis, J.V., Kulis, B., Jain, P., Sra, S., Dhillon, I.S.: Information-theoretic metric learning. In: Proceedings of the 24th international conference on Machine learning, pp 209–216 (2007)
45. Zhao, R., Ouyang, W., Wang, X.: Person re-identification by salience matching. In: Proceedings of the International Conference on Computer Vision, pp. 2528–2535 (2013)
46. Weinberger, K.Q., Saul, L.K.: Distance metric learning for large margin nearest neighbor classification. JMLR 10, 207–244 (2009)

47. Zheng, W.S., Gong, S., Xiang, T.: Person re-identification by probabilistic relative distance comparison. In: CVPR (2011)
48. Mignon, A., Jurie, F.: PCCA: A new approach for distance learning from sparse pairwise constraints. In: CVPR (2012)
49. Lisanti, G., Masi, I., Del Bimbo, A.: Matching people across camera views using kernel canonical correlation analysis. In: Proceedings of the International Conference on Distributed Smart Cameras Article, pp 1–6 (2014)
50. Prosser, B., Zheng, W.S., Gong, S. Xiang, T., Mary, Q.: Person re-identification by support vector ranking. In: Proceedings of the British Machine Vision Conference, pp. 1–11 (2010)
51. Li, Z., Chang, S., Liang, F., Huang, T.S., Cao, L., Smith, J.R.: Learning locally-adaptive decision functions for person verification. In: Proceedings of the IEEE Conference on Computer Vision and Pattern Recognition, pp. 3610–3617 (2013)
52. Zheng, L., Shen, L., Tian, L., Wang, S., Wang, J., Tian, Q.: Scalable person re-identification: A benchmark. In: Proceedings of the IEEE International Conference on Computer Vision, pp. 1116–1124 (2015)
53. Li, W., Zhao, R., Xiao, T., Wang, X.: DeepReID: Deep Filter Pairing Neural Network for Person Re-Identification. IEEE Conference on Computer Vision and Pattern Recognition (2014)
54. Zhao, L., Li, X., Zhuang, Y., Wang, J.: Deeply-learned part-aligned representations for person re-identification. In: IEEE international conference on computer vision, pp. 3219–3228 (2017)
55. Yu, H.-X., Zheng, W.-S., Wu, A., Guo, X., Gong, S., Lai, J.-H.: Unsupervised Person Re-identification by Soft Multilabel Learning. In: Proceedings of the IEEE International Conference on Computer Vision and Pattern Recognition (CVPR) (2019)
56. Li, M., Zhu, X., Gong, S.: Unsupervised Tracklet Person Re-Identification. IEEE Transactions on Pattern Analysis and Machine Intelligence 42(7) (2020)
57. Yu, H.-X., Wu, A., Zheng, W.-S.: Unsupervised Person Re-identification by Deep Asymmetric Metric Embedding. IEEE Transactions on Pattern Analysis and Machine Intelligence (2019)
58. Zhong, Z., Zheng, L., Cao, D., Li, S.: Re-ranking Person Re-identification with k-reciprocal Encoding. The IEEE Conference on Computer Vision and Pattern Recognition, pp. 318–1327 (2017)
59. Bazzani, L., Cristani, M., Murino, V.: Symmetry-driven accumulation of local features for human characterization and re-identification. Computer Vision and Image Understanding 117(2), 130–144 (2013)
60. Wu, L., Shen, C., van den Hengel, A.: PersonNet: Person Re-identification with Deep Convolutional Neural Networks. arXiv: 1601.07255v2 (2016)
61. Zhao, R., Ouyang, W., Wang, X.: Unsupervised salience learning for person re-identification. In: Proceedings of the IEEE Conference on Computer Vision and Pattern Recognition, pp. 3586–3593 (2013)
62. Rena, C-X., Xub, X-L., Leic, Z.: A Deep and Structured Metric Learning Method for Robust Person Re-Identification. Pattern Recognition 96 (2019)
63. Chen, J., Wang, Y., Qin, J., Liu, L., Shao, L.: Fast person re-identification via cross- camera semantic binary transformation. IEEE Conference on Computer Vision and Pattern Recognition, pp. 3873–3882 (2017)
64. Liu, F., Zhang, L.: View Confusion Feature Learning for Person Re-identification. IEEE International Conference on Computer Vision Foundation (2019)
65. Zhou, K., Yang, Y., Cavallaro, A., Xiang, T.: Learning Generalisable Omni Scale Representations for Person Re-Identification. IEEE Transactions on Pattern Analysis and Machine Intelligence (2021)

Automatic Segmentation of Handwritten Devanagari Word Documents Enabling Accurate Recognition

Mohammad Idrees Bhat[1]([⊠]), B. Sharada[2], Mohammad Imran[3], and Sk. Md Obaidullah[4]

[1] School of Computer Science, MIT-World Peace University, Pune, Maharashtra, India
idrees11@yahoo.com
[2] Department of Studies in Computer Science, University of Mysore, Mysuru, Karnataka, India
[3] NTT DATA Information Processing Services Private Limited, Bangalore, Karnataka, India
[4] Department of Computer Science and Engineering, Aliah University, West-Bengal, India

Abstract. In this paper, we propose different approaches for the segmentation of handwritten Devanagari word documents into constituent characters (or pseudo-characters). For accurate identification and segmentation of shiroreakha we exploited *ShiroreakhaNet* which is encoder-decoder based convolutional neural network. After, segmenting the shiroreakha structural patterns/properties are exploited for the segmentation of upper and lower modifiers. For the corroboration of the efficacy of the results, we collected dataset from different domains. Comparison is also performed with the state-of-the-art methods, and it was revealed that proposed approaches significantly perform better.

Keywords: Segmentation of handwritten Devanagari words · Pseudo-characters · Shiroreakha · ShireakhaNet · Upper and lower modifiers

1 Introduction

Varying individual handwritten styles, degenerative condition of scanned documents, script characteristics, cursiveness in handwriting, make segmentation of handwritten Devanagari words into its constituent characters (or *pseudo-characters*) a difficult task. In contrast, there exist number of successful attempts for printed documents thanks to their controlled nature of writing. In fact, segmentation of printed word documents with the help from *optical character recognition* (OCR) is widely believed as solved problem [1–3]. On the contrary, there are various issues that need to be looked in the direction of automatic segmentation of handwritten word documents, such that, accurate word/character recognition is achieved. Hence, it is widely considered as an open and challenging area of research.

Broadly speaking, the main aim in character segmentation is to decompose word documents into sequence of individual characters [4]. The degree of correctness/accurateness of segmentation governs the recognition success [5]. Despite such an importance, it is

© Springer Nature Switzerland AG 2022
R. Chbeir et al. (Eds.): MIKE 2021, LNAI 13119, pp. 78–88, 2022.
https://doi.org/10.1007/978-3-031-21517-9_8

striking, as we have come across a limited number of studies in this direction. For example, in [6], a structural approach is proposed for viewing the structural similarity of different characters, like for example, presence and exact location of vertical and joint bars in the characters. Later, segmentation is carried out in hierarchical order. However, most of the words are not segmented properly as writer frequently violates general conditions of script while writing. Continuing the structural patterns, horizontal and vertical pixel densities, etc., were exploited in [7]. In [8–10], almost all have exploited similar structural properties and image processing techniques for carrying out identification and segmentation of characters.

Notwithstanding, all the stated approaches segmentation of handwritten Devanagari words into its constituent characters still remains a challenging problem. More often words are not segmented into characters accurately. It is mainly due to the script characteristics, for example, shiroreakha is not identified accurately instead essential character information gets identified and subsequently segmented/removed. Moreover, identification and segmentation of modifiers (upper and lower), complexities in the middle zone further complicate the problem. With this motivation, in this paper, we have come up with some interesting structural based approaches for the accurate segmentation. Which have demonstrated the substantial recognition performance. The rest of the paper is divided into four sections. Section 2 gives brief description about Devanagari script. The proposed methodology for segmentation of characters is described in Sect. 3. In Sect. 4 extensive experimentation is given. Finally, conclusion and future work is drawn in Sect. 5, respectively.

2   Brief Overview of Devanagari Script

Devanagari script is written from left-to-right and contains 13 and 34 vowels and consonants, respectively. As can be observed from Fig. 1 handwritten words are written in three zones, upper, middle and lower and contain a headline or *shiroreakha*. It is to be noted that vowels can be placed at the left, right (or both), top, and bottom of the constant. A *half-form* of the character is written by removing the vertical line of the character. Also, an *orthographic shape* (compound character), gets formed when a consonant is merged with one or more consonants. Note, Devanagari is cursive in hand writing (Fig. 1). Overall, stated artefacts make segmentation of handwritten Devanagari words into individual characters a challenging problem. For further details about the characteristics/properties of Devanagari script, we refer readers to [11–13].

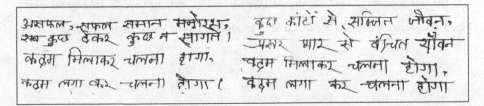

Fig. 1. Illustration of a Devanagari words.

3   Proposed Method

Handwritten word document/image is first resized from original $m \times n$ dimension to a new $k \times l$ dimension ($k = l = 256$) for coherence and simplicity. Next, resized images are filtered with median filtering technique (with window size 3×3). The processed word document thus obtained is passed through the *ShiroreakhaNet*, an encoder-decoder based convolutional neural network architecture which accurately identifies and segments the pixels of shiroreakha, as discussed in [14]. This step is followed by the segmentation of upper and lower modifiers. For their segmentation, we exploited structural patterns of the script. Finally, each word is decomposed into constituent characters (pseudo-characters). The proposed method is illustrated in Fig. 2.

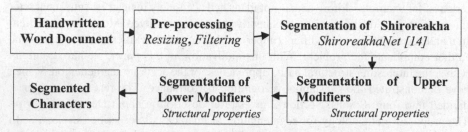

Fig. 2. Overview of the proposed methodology.

3.1   Identification and Segmentation of Shiroreakha [14]

In *ShiroreakhaNet* pixels belonging to background, character, and shiroreakha get accurately identified and finally segmented. Since the *ShiroreakhaNet* is generalizable and adaptable, therefore, with the concept of transfer learning [15], we exploited the *ShiroreakhaNet* [14].

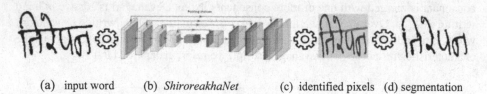

(a)   input word   (b) *ShiroreakhaNet*   (c) identified pixels   (d) segmentation

Fig. 3. Segmentation of Shiroreakha from a handwritten word image through *ShiroreakhaNet*.

3.2 Segmentation of Upper Modifiers

Section 3.1 follows the computation of vertical density of foreground pixels for each column of a word document. Cut-points/segmentation points are those columns which contain zero foreground pixels. This approach is *de-facto* in this direction. Afterwards, upper modifiers are segmented separately. For that, connected component labelling [16], is carried out and each individual component containing upper modifier is extracted. We know that the upper modifiers are classified into single or double touching, that is, number of times modifier makes contact with shiroreakha (shown in Fig. 4). Hence, for each class separate techniques were designed. Figure 7 shows the complete flowchart, however, following are the main steps:

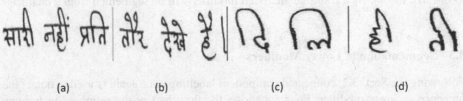

     (a)   (b)   (c)   (d)

Fig. 4. Different types of upper modifiers: (a) double touching (b) single touching (c) left modifier (d) right modifier

For double touching modifier:

Step 1. Skeletonized the copy of the component (C_s) containing the modifier after keeping original component (C_o) intact.

Step 2. Divide the C_s into two halves vertically. Further, divide each half horizontally into equal parts. Identify each component as: top left part I_1, left bottom I_2, top right I_3, and right bottom I_4.

Step 3. Check whether the modifier is left or right to the character in C_o. Figure 4 illustrates the left and right double touching upper modifiers.

Step 4. Start from the bottom of I_3, scan each row and stop when encountering first occurrence of foreground pixel. Record the $(x, y)-$ coordinates of that foreground pixel.

Step 5. Up to four consecutive positions, down to the $x-$ coordinate, for all columns (starting from first column to the less than or equal to the size $C_o/2$ columns) in I_3 of C_o, invert any foreground pixel into background. As a result right modifier will be segmented from a character (Fig. 5).

Step 6. Repeat step 4 for left modifier also. Now invert four consecutive positions of $x-$ coordinate, for all columns (starting from the size of $C_o/2$ to the size of C_o). As a result left modifier will be segmented from a character (Fig. 5).

For single touching modifier:

Step 1. Repeat steps 1–3 of double touching modifier technique.
Step 2. Start from the top of I_3, scan each row and stop when encountering first occurrence of foreground pixel. Record the $(x, y)-$ coordinates of that foreground pixel.
Step 3. Invert up to four consecutive positions, down to the $x-$ coordinate after adding the size of $C_o/2$ to each. Invert corresponding columns (starting from the size of $C_o/2$ to the size of C_o). As a result single touch modifier will be segmented from a character (Fig. 5).

3.3 Segmentation of Lower Modifiers

Following the Sect. 3.2, connected component labelling once again is used to detect the presence of lower modifier. Figure 8 shows the flowchart of the proposed technique, however, following are the main steps:

Step 1. Label the connected components in a word image.
Step 2. Compute the length L from each connected component and detect the component with lower modifier (Fig. 6).
Step 3. Compute the maximum height from L and delete it.
Step 4. Take minimum height L_{min} among component heights in L. Segment out each component with L_{min} height and its corresponding width from the word image As a result lower modifier will be segmented from a character (Fig. 6).

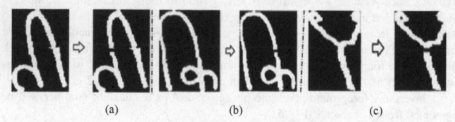

(a)   (b)   (c)

Fig. 5. Segmentation of upper modifier: (a) right modifier (b) left modifier (c) combined single touching modifier

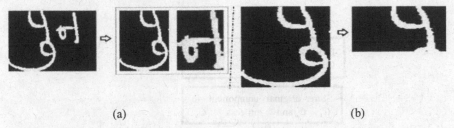

(a)   (b)

Fig. 6. (a) Illustration of connected component labeling (b) segmentation of lower modifier

4 Experimentation

4.1 Dataset

We manually selected Devanagari words from legal amounts (amount written in words on bank Cheques) [17], and document dataset [18], respectively. Legal amounts are written in both Hindi and Marathi languages. And, is collected from specially bank forms, which contains different boxes where each writer has to write posible lexicon of legal amounts by following a specific order. Document dataset is a small in-house dataset created towards the graph based recognition of handwritten Devanagari words. It comprises of word images from 150 documents. Fifty writers were used to write these documents. Note, the words were selected such that all the complexities (see Sect. 1) are reflected in the collected dataset.

4.2 Experimental Results

We have divided the experiments into three stages. The first experiment consists of segmentation of shiroreakha, next segmentation of upper modifiers, and finally segmentation of lower modifiers. Note, accuracy is determined through visual inspection of output, as it is generally followed in literature [6]. That is, *"how one can easily recognize/identify segmented constituents (upper modifiers, lower modifiers, and character)."* We calculated accuracy by using Eq. (1).

$$Accuracy = \frac{Number\ of\ constituents\ identified\ in\ a\ word\ after\ segmentation}{Total\ number\ of\ constituents\ present\ in\ a\ word} \times 100$$

(1)

4.2.1 Segmentation of Shiroreakha

Table 1 shows the recognition accuracy obtained and Fig. 9 shows some resulting word samples, first row shows the input word sample and second shows the output.

4.2.2 Segmentation of Upper Modifiers

Experimental results are shown in Table 2 and Fig. 10 shows the some segmented samples. We can observe from the Fig. 10 most of the modifiers are segmented from the middle zone stroke to which they are attached. It is worth to state here, no joining has been performed to attach modifiers with their corresponding character.

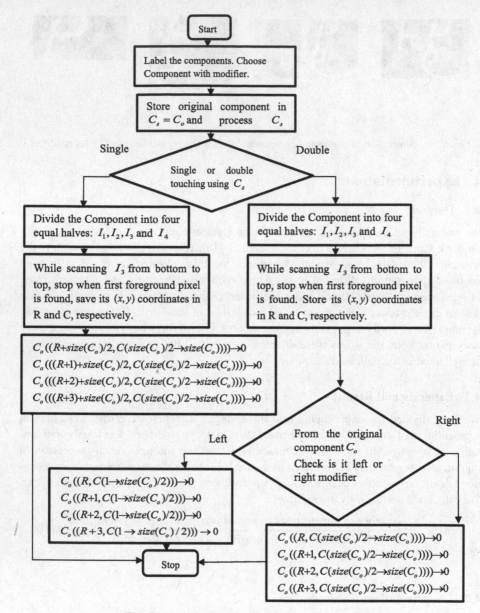

Fig. 7. Flow chart for upper modifier segmentation

4.2.3 Segmentation of Upper Modifiers

Lower modifiers are segmented at the last from the word document. The experimental results are given in Table 3 and some samples are illustrated in Fig. 11. It is observed that almost all lower modifiers are segmented correctly. However, in some cases, small parts of lower modifiers remain attached to corresponding character.

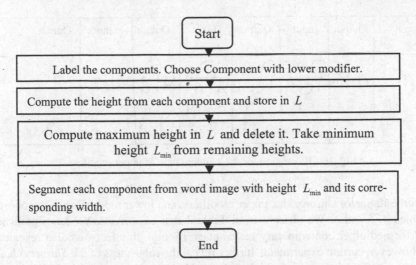

Fig. 8. Flowchart for lower modifier segmentation

Table I. Segmentation of shiroreakha with *ShiroreakhaNet*.

Method	Number of words	Shiroreakha segmented correctly	Accuracy
ShiroreakhaNet [14]	1000	950	95%

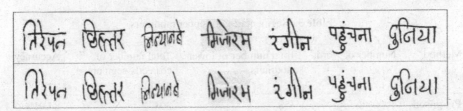

Fig. 9. Segmentation of shiroreakha

Table 2. Segmentation of Upper Modifiers

Method	Number of words	Total number of upper modifiers	Total number of correctly segmented modifiers	Accuracy
Proposed	1000	536	460	85.82%

4.2.4 Comparative Study

Due to the non-availability of datasets from previous works [6, 19, 20]. Therefore, we decided to implement these techniques on our collected dataset. The experimental results

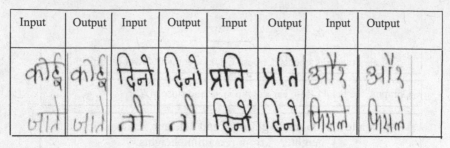

Fig. 10. Illustration for the segmentation of upper modifiers

for segmentation of shiroreakha, upper modifiers, and lower modifiers are corroborated in Tables 4, 5 and 6. We observe from Table 4 that *ShiroreakhaNet* has significantly outperformed other contemporary techniques. Though, this fact was also revealed in [14], however, current experiments further reveal the robustness of the *ShiroreakhaNet*. (Note, the techniques against which *ShiroreakhaNet* are compared in [14], are more or less similar proposed in [6, 19, 20]). This is mainly due to the generalizability and adaptability of *ShiroreakhaNet*. Moreover, for segmentation of upper and lower modifiers proposed techniques have achieved acceptable accuracies against techniques proposed in [6, 19]. However, technique proposed in [7], has been marginally outperformed. The overall accuracy of our proposed method is 88.57% which is better than the other three techniques [6, 19, 20], as shown in Table 5 and Table 6. Note, over and under segmentation have been treated as incorrect segmentation while measuring the accuracies.

Table 3. Segmentation of upper modifiers

Method	Number of words	Total number of lower modifiers	Total number of correctly segmented modifiers	Accuracy
Proposed	1000	358	304	84.91%

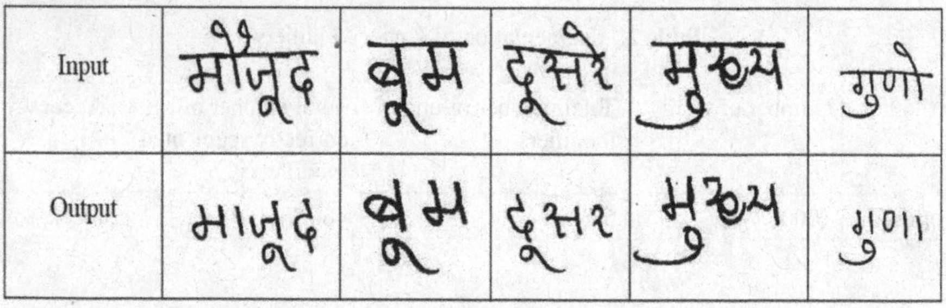

Input	मौजूद	बूम	दूसरे	मुख्य	गुणो
Output	माजूद	बूम	दूसर२	मुख्य	गुण||

Fig. 11. Illustration for segmentation of lower modifiers

Table 4. Comparison of segmentation of shiroreakha

Method	Number of words	Shiroreakha segmented correctly	Accuracy
[19]	1000	446	44.60%
[6]		737	73.70%
[7]		861	86.10%
ShiroreakhaNet [14]		956	**95.60%**

Table 5. Comparison of segmentation of upper modifiers

Method	Number of words	Total number of upper modifiers	Total number of correctly segmented modifiers	Accuracy
[19]	1000	536	230	42.91%
[6]			391	72.94%
[7]			440	82.08%
Proposed			460	**85.82%**

Table 6. Comparison of segmentation of lower modifiers

Method	Number of words	Total number of lower modifiers	Total number of correctly segmented modifiers	Accuracy
[19]	1000	358	146	40.78%
[6]			264	73.74%
[7]			289	80.72%
Proposed			304	**84.91**

5   Conclusion

In this paper, some interesting techniques are proposed towards segmentation of handwritten word documents into different constituents/characters/pseudo-characters. The proposed method has exploited the *ShiroreakhaNet* [14] and structural patterns/properties of the script, for segmentation of word documents into its individual characters. The experimentation is carried on 1000 words collected from two diverse datasets. From the experimentation it was revealed that proposed approaches substantially outperforms two out of three state-of-the-art methods. However, scalability of these approaches needs to be ascertained on large dataset. Further, there is every possibility that characters might be touched and overlapped. Addressing these issues are rewarding avenues to deal with.

References

1. Bhunia, A.K., Roy, P.P., Sain, A., Pal, U.: Zone-based keyword spotting in Bangla and Devanagari documents. Multimedia Tools and Applications **79**(37–38), 27365–27389 (2020). https://doi.org/10.1007/s11042-019-08442-y
2. Bhunia, A.K., Mukherjee, S., Sain, A., Bhunia, A.K., Roy, P.P., Pal, U.: Indic handwritten script identification using offline-online multi-modal deep network. Information Fusion. **57**, 1–14 (2020)
3. Bhat, M.I., Sharada, B.: Automatic recognition of legal amounts on indian bank cheques: a fusion based approach at feature and decision level. Int. J. Comp. Vision and Image Procesing **10**(4), 57–73
4. Casey, R.G., Casey, R.G., Lecolinet, E., Lecolinet, E.: A survey of methods and strategies in character segmentation. Analysis. IEEE Transactions on Pattern Analysis and Machine Intelligence **18**, 690–706 (1996)
5. Bortolozzi, F., Souza, A. De B., Jr, Oliveira, L.S., Morita, M.: Recent advances in handwriting recognition. IEEE Trans. Pattern Anal. Mach. Intell. **22**, 38–62 (2004)
6. Murthy Ramana, O.V.: An approach to offline handwritten Devanagari word segmentation. Int. J. Comput. Appl. Technol. **44**, 284–292 (2012)
7. Bag, S., Krishna, A.: Character segmentation of Hindi unconstrained handwritten words. Lect. Notes Comput. Sci. **9448**, 247–260 (2015)
8. Bhujade, M.V.G., Meshram, M.C.M.: A technique for segmentation of handwritten Hindi text. Int. J. Eng. Res. technoogy. **3**, 1491–1495 (2014)
9. Kohli, M., Kumar, S.: Pre-segmentation in offline handwritten words. Infocomp J. Comput. Sci. **18**, 48–53 (2019)
10. Ramteke, A.S., Rane, M.E.: Offline handwritten Devanagari script segmentation. Int. J. Sci. Technol. Res. **1**, 142–145 (2012)
11. Pal, U., Chaudhuri, B.B.: Indian script character recognition: a survey. Pattern Recognit. **37**, 1887–1899 (2004)
12. Jayadevan, R., Kolhe, S.R., Patil, P.M., Pal, U.: Offline recognition of Devanagari script: a survey. IEEE Trans. Syst. Man Cybern. Part C Appl. Rev. **41**, 782–796 (2011)
13. Bag, S., Harit, G.: A survey on optical character recognition for Bangla and Devanagari scripts. Sadhana - Acad. Proc. Eng. Sci. **38**, 133–168 (2013)
14. Bhat, M.I., Sharada, B., Obaidullah, S.M., and Imran, M.: Towards accurate identification and removal of shirorekha from off-line handwritten devanagari word documents. Proc. ICFHR 2020-September, pp. 234–239 (2020)
15. Kaya, A., Keceli, A.S., Catal, C., Yalic, H.Y., Temucin, H., Tekinerdogan, B.: Analysis of transfer learning for deep neural network based plant classification models. Comput. Electron. Agric. **158**, 20–29 (2019)
16. Gonzalez, R.C., Woods, R.E.: Digital image processing, https://books.google.com/books?id=lDojQwAACAAJ&pgis=1 (2008)
17. Jayadevan, R., Kolhe, S.R., Patil, P.M., Pal, U.: Database development and recognition of handwritten Devanagari legal amount words, pp. 304–308. Int. Conf. Doc. Anal. Recognition, Beijing (2011)
18. Malik, L.: A graph based approach for handwritten Devanagari word recognition. Fifth Int. Conf. Emerg. Trends Eng. Technol. **1**, 2012–42 (2012)
19. Bansal, V., Sinha, R.M.K.: Segmentation of touching and fused Devanagari characters. Pattern Recogn. **35**, 875–893 (2002)
20. Bag, S., Krishna, A.: Character segmentation of Hindi unconstrained handwritten words. Proc. 17th Int. Work. Comb. image Anal. **9448**, 247–260 (2015)

A Comparative Experimental Investigation on Machine Learning Based Air Quality Prediction System (MLBAQPS): Linear Regression, Random Forest, AdaBooster Approaches

G. Sujatha$^{(\boxtimes)}$ and P. Vasantha Kumari

Research Department of Computer Science, Sri Meenakshi Govt. Arts College for Women (A), Madurai, India
sujisekar05@gamil.com

Abstract. Air is the greatest critical expected possessions for the endurance plus subsistence for the complete life on this globe. All forms of the life comprising plant life plus, human life and wildlife depending on air for their basic survival. Thus, all breathing creatures require better excellence of air which is free of dangerous fumes for continuing their lives. Nowadays Air is polluted by the human activities, industrialization and urbanization. Investigative and protecting air quality in this earth has become one of the fundamental activities for every human in many industrial and urban areas today. Due to serious health concerns, atmospheric pollution has become a main source of premature mortality among general public by cause millions of deaths each year (WHO, 2014). Air Quality prediction used to alert the people about the air quality dreadful conditions, and it's health effects and it's also support environmentalist and Government to frame air quality standards and regulations based on issues of noxious and pathogenic air exposure and health-related issues for human welfare. Air Quality Index (AQI) is a measure of air pollution level. Predicting air pollution with AQI is one the major challenging area of Research. Machine Learning (ML) is used to predict the AQI. Machine Learning is a scientific approach to solve certain tasks and predict the value using techniques and Algorithms such as Supervised Learning (SL), Semi Supervised Learning (SSL), and Un Supervised Learning (USL). ML can be applied in many areas like web search, spam filter, credit scoring, fraud detection, computer design, recognition of network intruders or cruel insiders working towards a data dishonoring, etc., Machine Learning algorithms provide various methods to forecasting the air pollution levels. Three major Machine Learning algorithms are applied to predict the air quality and analyze these results to conclude with the comparison.

Keywords: Air Quality Index (AQI) · Machine Learning (ML) · Linear Regression (LR) · AdaBooster Regression · Random Forest Regression (RF) · Air pollution

© Springer Nature Switzerland AG 2022
R. Chbeir et al. (Eds.): MIKE 2021, LNAI 13119, pp. 89–101, 2022.
https://doi.org/10.1007/978-3-031-21517-9_9

1   Introduction

Air pollution is an important risk factor for a number of pollution related diseases, including respiratory infections, heart disease, and stroke, lung cancer [1]. The human health effects of poor air quality are far reaching, but mainly affect the body's lungs system and the cardio vascular system.. The increasing populace, vehicles and productions are poisoning all the air at an alarming rate. Dissimilar elements are related with metropolitan air contamination. Air excellence sensor devices extent the attentions of particles that have the sources of pollution and create the hazardous effects during or after the gulp of air by human being. Particles like PM2.5, CO, NO_2, NO etc. disturbs the quality of air. Automobiles release enormous amounts of nitrogen oxides, carbon oxides, hydrocarbons and particulates when burning petrol and diesel.

1.1   Air Pollutants

Air pollution is resolute as the presence of pollutants in the air in huge amount for long periods. It can either be solid particles, liquid. Air is polluted from various ways. The sources of air pollutants are listed in the table [2]. Droplets, or gases, which are classified into the following (Fig. 1) (Table 1):

Table 1.   Sources of air pollutants

Air pollutant Factors	Sources of Pollutant
Sulphur Dioxide (SO_2)	Very dangerous due to the nasty odour and it reacted quickly with the suspended particles in air to form the harmful acids which may arises the acid rain
Nitrogen Dioxide (NO_2)	Nitrogen oxides emitted from nitrogen compound from the fuel
Ozone (O_3)	Ozone is a secondary pollutant and it is formed reactions between pollutant emitted from industrial and automobile sources
Carbon Monoxide (CO)	Carbon monoxide (CO) produces by incomplete combustion of fuel such as natural gas, coal or wood. A major source of carbon monoxide is motor vehicle
Particulate Matter (PM) of less than 2.5 microns size (PM2.5)	It is emitted either from the combustion of solid and liquid fuels. It has a diameter of less than 2.5 μm
PM of less than 10 micron size (PM10)	PM_{10} is inhalable particles, with diameters that are generally 10 μm and smaller. It is emitted directly from construction sites, unpaved roads, fields, smokestacks or fires

(continued)

Table 1. (*continued*)

Air pollutant Factors	Sources of Pollutant
Lead (Pb)	Lead (Pb) is one of pollutant of air. It is emitted from ore and metals processing, piston-engine aircraft operating on leaded aviation fuel
Ammonia (NH_3)	Ammonia is one of the industrial chemical pollution in air and the same is produced by decomposition of plant,animal wastes
Benzo(a)Pyrene (BaP)	It is emitted from residential wood burning. It is also found in coal tar, in vehicle exhaust fumes (especially from diesel engines), in all burn resultant from the combustion of untreated material (including cigarette smoke), and in charbroiled food
Benzene (C_6H_6)	Benzene melt only slightly in water and will drift on top of water
Arsenic (As)	A lot of the arsenic in the environment comes from high-temperature processes such as coal-fired power plants, on fire vegetation and volcanic activity. It is released into the surroundings for the most part as arsenic trioxide where it adheres willingly onto the outside of particles
Nickel (Ni)	It presence in the air from the combustion of coal, diesel oil and fuel oil, and the burning of waste and sewage

- Primary Pollutants
  Primary Pollutants are emitted from the source directly to the atmosphere and sources can be either sandstorms or human-related, such as industry and motor vehicle emission. The common primary pollutants are sulphur dioxide (SO_2), particulate matter (PM), nitrogen dioxide (NO_2), and carbon monoxide (CO) [2].

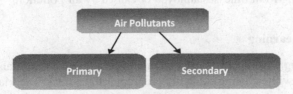

Fig. 1. Air pollutant classification

- Secondary Pollutants
  The air pollutants are formed in the atmosphere, resulting from the chemical or physical interactions between primary pollutants. Photochemical oxidants and resulting particulate matter are the major examples of secondary pollutants.

1.2 AQI (Air Quality Index)

AQI is the measure of determining Air Pollution Index which is the indicator of Pollution level of Air. The air pollution factors are used as parameters to calculate the AQI. Based on this value the AQI is ranked by six types [4, 14].These ranks are listed in Table 2.

Table 2. AQI rank categories

Rank	Categeory	Color	AQI Value	Precautionary Message
1	Good	Green	0-50	None
2	Satisfactory	Yellow	51-100	Unusually responsive people should consider reducing long-standing or heavy exertion.
3	Moderately Polluted	Orange	101-150	People with respiratory or heart disease, the elderly and children should limit prolonged exertion
4	Poor	Red	151-200	People with respiratory or heart disease, the elderly and children should avoid long -standing exertion; everyone else should limit pr o-longed exertion
5	Very poor	Purple	201-300	Active children and adults, and people with respiratory disease, such as asthma, should avoid all outside physical exertion; everyone else, especially children, should limit outdoor exertion
6	Serve	Maroon	301-500	Everyone should avoid any outdoor exertion; people with respiratory or heart disease, the old and children should remain at home.

Objectives of Air Quality Index (AQI)

- AQI used to comparing air quality environment at different locations and cities.
- It identifying damaged standards and inadequate monitoring programmes.
- AQI helps in analyzing the change in air quality (upgrading or degradation).
- AQI informs the public about environmental conditions. It is especially useful for people suffering from illnesses annoyed or caused by air pollution.

1.3 Machine Learning

Machine Learning is a branch of Artificial Intelligence. Its goal is to enable the computer to learn by itself without being explicitly programmed the rules. The Machine Learning algorithm can identify and learn underlying patterns in observed data to model and predict the world [3].

There are three kinds of Machine Learning techniques (Fig. 2) Reinforcement Learning, Unsupervised Learning, and Supervised Learning. In Reinforcement Learning, the algorithm receives feedback based on performance as it navigates its problem space. Tasks such as playing a game or driving a car are examples where Reinforcement Learning is suitable.

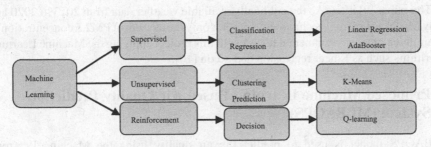

Fig. 2. Machine learning methods

Unsupervised Learning is an approach that learns from data that is unlabeled or classified. Instead of responding to feedback as in Reinforcement Learning, unsupervised Learning identifies shared attributes and characteristics from association problems, which try to describe parts of the data, and clustering problems, that seek to identify natural groupings. In supervised Learning, the algorithm attempts to learn from informative examples of labeled data [7]. Such algorithms can be described as a data- driven approach, where historical data is used for predictions of the future.In the paper, Air Quality is predicted with the help of three Machine Learning Algorithms and the results are compared against different parameters.

2   Related Work

In recent years there has been a number of Machine Learning methods proposed for solving air pollution prediction problems. An overview of various air pollution forecasting algorithms is provided in [5]. The paper explains and reviewed theory and applications of multiple predictive models as well as further compared advantages and disadvantages among models.

In [6] it is analyzed various different regression analysis techniques for accurate results. It explains the relationship between a dependent and independent variable. This technique is used for forecasting or predicting, time series modeling, and finding the causal effect relationship of the variables of the model. Regression analysis is used to analyzing and modeling data. There are different kinds of regression techniques used to make predictions namely Linear regression, Support vector regression, Decision tree regression and Lasso regression. The system [7] has used to predict the next day pollution level using Linear Regression and Multilayer Perceptron (ANN) Protocol. Based on basic parameters and analyzing pollution details and forecast future pollution for next day. Time Series Analysis is used for recognition of future data points and air pollution prediction.

In this paper [8] to predict the air pollution using Supervised Machine Learning approach considers three Machine Learning algorithms such as LR, Adabooster and RF. The absorption of air pollutants in ambient air is governed by the meteorological parameters such as atmospheric wind speed, wind direction, relative humidity, and temperature, using these parameters to predict the Air Quality Index. It is used to measure the quality of air.

The paper [9] is to take the publically available weather data from 2013 to 2020 and apply Machine Learning techniques to predict only the amount of PM2.5 concentration in the air given other environmental features. In this project applying the Machine Learning algorithms such as Linear regression and so on [10].

3 Proposed Machine Learning Based Air Quality Prediction System (MLBAQPS)

MLBAQPS model is used to predict the air quality using the Machine Learning approaches using air pollutant data without metrological data whereas the previous researches using the both metrological and air pollutant data to predict the air quality, but in this model used only the air pollutant data. The desired data is downloaded from kaggle website. This model is used to comparing and finding best suitable algorithm for AQI prediction.

The Proposed Machine Learning Based Air Quality Prediction System (MLBAQPS) is designed for purpose of comparing initial Machine learning algorithm. It is a Layered Architecture Model (Fig. 3). Based on the functionalities MLBAQPS is divided in to 5 layers. The bottom most layer is data layer which is used to collect data from the kaggle database. The next layer is data Filtering layer that will clean data by removing the missing value and unknown value. The third layer is Data processing layer used to split the data into testing and training data set. The next layer is Prediction layer is used to predict the AQI value using Machine Learning layer used to split the data into testing and training data set. The next layer is Prediction layer is used to predict the AQI value using Machine Learning methods. This layer is heart of this model. The top most layer is output layer is take the predicted AQI value.

3.1 Data Layer

The bottom most layer of MLBAQPS architecture is Data layer. This layer used to collect the data from the kaggle database. Kaggle is interactive online platform provides hundreds of databases and tutorials. The data is structured data type. This data is downloaded by.csv(comma separated value) format.

Dataset Description
The dataset consists of around 29532 records and 13 attributes. The attributes are 1.PM2.5, 2.PM10, 3.NO, 4.NO$_2$, 5.NOX, 6.NH$_3$, 7.CO, 8.SO3 9.O3, 10.Benzene 11.Toluene, 12.Xylene, 13.AQI.

3.2 Data Filtering Layer

The Second layer of MLBAQPS model is data Filtering layer. Data filtering is important step in Machine Learning model. The data set is having a large number of noisy data. The unknown data and missing data is removed by using the method pandas in python.Pandas packages used to eliminate rows and columns with Null/NaN values. By default, this function returns a new Data Frame and the source Data Frame remains unaffected.

3.3 Data Processing Layer

Data Processing layer is used to split the dataset into training dataset and testing dataset for Machine Learning Algorithms. The first 13 parameters of the dataset are considered as input parameters X and the last parameter is taken as the output parameter Y. Model takes the input as x and given the output to y. The dataset is split by training and testing using the method train_test_split in sklearn. The dataset spit by 0.5% of data. One is testing another one is training data.After removing the NaN value the data set consists only 6236 records. These records are used to build the model.

3.4 Prediction Layer

The heart of MLBAQPS is prediction layer. This layer is used to predict the desired air pollution level.Machine Leaning methods are used to predict the (Fig. 4).

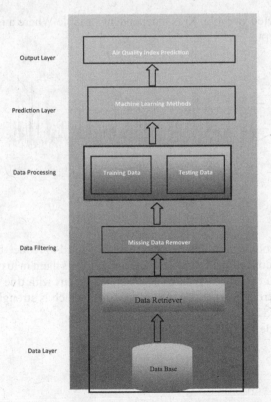

Fig. 3. Layered Architecture model of AQI prediction

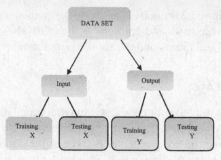

Fig. 4. Dataset Splitting

Linear Regression

Linear Regression is one of the Machine Learning algorithms [11]. It is Used for prediction. In this paper Linear regression is used to predict the air quality index value.

Linear Regression Model:

The equation of the linear regression is given (Fig. 5)

$$Yp = Xi(a + b), \tag{1}$$

where Yp is predicted variable, Xi is independent variable Where a is slope of the line and b is the intercept.

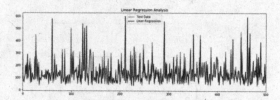

Fig. 5. Linear Regression Analysis

AdaBooster

Boosting is a method in ensemble Learning which is used to reduce bias and variance. Boosting creates a collection of weak learners and converts them in to one strong learner. A weak learner is a classifier which is hardly simultaneous with true classification. On the other hand, a strong learner is a type of classifier which is strongly correlated with true classification [12].

The pseudo code for AdaBoost Algorithm

Input: Data set $\mathcal{D} = \{(\boldsymbol{x}_1, y_1), (\boldsymbol{x}_2, y_2), \cdots, (\boldsymbol{x}_m, y_m)\}$;
Base learning algorithm $\mathcal{L}$;
Number of learning rounds T.

Process:

$D_1(i) = 1/m.$

for $t = 1, \cdots, T$:

$h_t = \mathcal{L}(\mathcal{D}, D_t);$

$\epsilon_t = \text{Pr}_{i \sim D_i}[h_t(\boldsymbol{x}_i \neq y_i)];$

$\alpha_t = \frac{1}{2}\ln \frac{1-\epsilon_t}{\epsilon_t}$;

$D_{t+1}(i) = \frac{D_t(i)}{Z_t} \times \begin{matrix} \exp(-\alpha_t) & \text{if } h_t(\boldsymbol{x}_i) = y_i \\ \exp(\alpha_t) & \text{if } h_t(\boldsymbol{x}_i) \neq y_i \end{matrix}$

$\qquad\quad = \frac{D_t(i)\exp(-\alpha_t y_i h_t(\boldsymbol{x}_i))}{Z_t}$

end.

Output: $H(\boldsymbol{x}) = \text{sign}(f(\boldsymbol{x})) = \text{sign} \left(\sum_{t=1}^{T} \alpha_t h_t(\boldsymbol{x}) \right)$

The comparison of testing data and predicted AQI of Adabooster regression model is shown in Fig. 6 (Fig. 7).

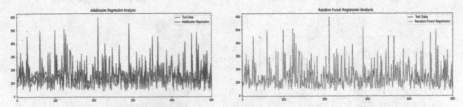

Fig. 6. ADA Booster Regression   **Fig. 7.** Random Forest Regression analysis

Random Forest is defined as a set of decision trees to do regression and classification. Classification is used to find out the majority selection. This algorithm is more accurate, robust, and can manage a variety of data such as binary data, categorical data, and continuous data.

Random Forest is nothing but multiple decision trees. The training data is subjected to sampling and based on attribute sampling different decision trees are constructed by applying the Random Forest [13]. In Fig. 8. Shown the comparison of testing data and predicted AQI of Random Forest regression model.

3.5 Output Layer

The top most layer of MLBAQPS is output layer. In this layer take the predicted AQI value Y. Compared the predicted and testing AQI based on this accuracy is generated.

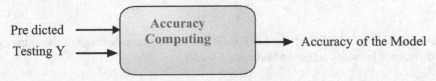

Fig. 8. Accuracy Checking.

4 Results and Discussion

MLBAQPS model the training data set used to fit the Model and test dataset used to predict the AQI and calculate the accuracy of the model. The data set [15] is cleaned (Fig. 9 and Fig. 10).

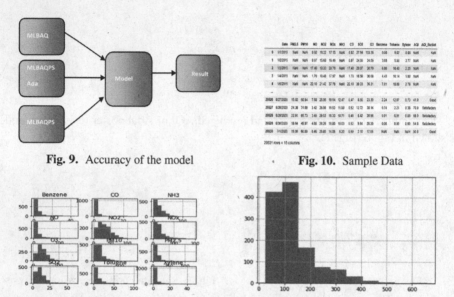

Fig. 9. Accuracy of the model

Fig. 10. Sample Data

Fig. 11. Training X dataset for Machine Learning

Fig. 12. Training Y (AQI) Value.

The same dataset is used for various algorithms to predict the air quality. The 12 input parameters of air pollution dataset value are shown Fig. 11. The Training AQI value is shown on Fig. 12. AQI is predicted from the testing dataset and the accuracy of the model is analyzed using various performance estimators" like mean square error, root mean square error, absolute mean error etc.

ROOT MEAN SQUARE ERROR (RMSE):
Root Mean Square Error (RMSE) is a standard deviation of residuals (prediction errors). RMSE measures, how to extent these residuals. RMSE value observed for this model is: 22.1

$$RMSE = \sqrt{\frac{1}{n} \sum_{j=1}^{n} (yj - \hat{y}j)^2} \qquad (2)$$

MEAN SQUARED ERROR: The MSE isdefined as Mean or Average of the square of the difference between actual and estimated values.

$$MSE = \frac{1}{n} \sum_{j=1}^{n} (y_i - \hat{y}_i)^2 \qquad (3)$$

MEAN ABSOLUTE ERROR: The mean absolute error (MAE) is very simple regression error metric to comprehend. To calculate the residual for every data point, taking only the absolute value of each, so that negative and positive residuals do not cancel out then take the average of all these residuals (Table 3 and Fig. 13).

$$MAE = \frac{1}{N} \sum_{i=1}^{N} |y_i - \hat{y}_i| \tag{4}$$

Table 3. Accuracy Comparison of Machine Learning models

ML Algorithms	MAE	MSE	RMSE	Accuracy
Linear Regression	18	770	27.75	91.17%
AdaBooster	34	1542	39.27	82.2%
Random Forest	14	489	22.1	97.9%

Observed value of Mean absolute error is 14.

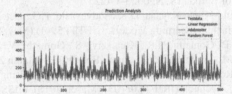

Fig. 13. Prediction Analysis of Machine Learning Methods

Efficiency of the given model is analyzed using 4 performance estimators they are:
Observations: Over fitting is less in Random Forest because many decision trees are involved

- Less error in classification in RF: RF gives Better performance compared to AdaBooster, Linear Regression.
- Less Variance compared in RF. RF is a highly flexible Algorithm.
- The Minimum time is required for pre- processing the data.
- RF is a very stable algorithm.

Based on this comparison Random Forest Regression is better than other 2 Regression.

5 Conclusion

Air pollution is one of the most important problems in this world and its prevention is stable scientific challenges. The AQI used to inform the quality of the air and its health effects. The AQI is very important to sensitive people. The regression algorithm is simple and easy to predict the air quality in Machine Learning algorithm and this paper used basic three regressions only. Comparison of these regressions, the Random Forest regression is better than others. In future the advanced ANN algorithm used to predict the AQI prediction. Concentration of air pollutants is governed by the meteorological parameters such as atmospheric wind speed, wind direction, relative humidity, and temperature must be considered in future to reduce the error rate and increase the accuracy of the model and the AQI is classified into several categories based on this predicted value.

References

1. 7Million Premature Deaths Annually Linked to Air Pollution. Accessed: 27 April 2019. Available: https://www.who.int/phe/eNews_63.pdf
2. Sujatha, G., Vasantha Kumari, P., Vanathi, S.: Machine Learning Approaches for Air Quality Prediction – A Review. In: International Conference on Digital Transformation (AC-ICDT"20), ISBN-978-93-87865-29-7
3. Osisanwo, F.Y., Akinsola, J.E.T., Awodele, O., Hinmikaiye, J.O., Olakanmi, O., Akinjobi, J.: Supervised Machine Learning Algorithms: Classification and Comparison International Journal of Computer Trends and Technology (IJCTT) **48**(3) (June 2017)
4. Sujatha, G., Vasantha Kumari, P.: Overview on air quality index (AQI). Journal of the Maharaja Sayajirao University of Baroda ISSN: 0025–0422
5. Aditya, C.R., Deshmukh, C.R., Nayana, D.K., Vidyavastu, P.G.: Detection and Prediction of Air Pollution using Machine Learning Models. (IJETT) **59**(4) (May 2018)
6. Aarthi, A., Gayathri, P., Gomathi, N.R., Kalaiselvi, S., Gomathi, V.: Air Quality Prediction Through Regression Model. Int. J. Sci. Technol. Res. **9**(03), (March 2020)
7. RuchiRaturi, J.R.P.: Recognition Of Future Air Quality Index Using Artificial Neural Network. Int. Res. J. Eng. Technol. (IRJET) **05**(03) (March 2018). e-ISSN: 2395-0056 p-ISSN: 2395-0072
8. Madhuri, V.M., Samyama Gunjal, G.H., Kamalapurkar, S.: Air Pollution Prediction Using Machine Learning Supervised Learning Approach. Int. J. Sci. Technol. Res. **9**(04), (April 2020). ISSN 2277-8616
9. Ruchita Nehete, D.D.: Patil air quality prediction using machine learning. Int. J. Creative Res. Thoughts (IJCRT) © 2021 IJCRT **9**(6) (June 2021). ISSN: 2320–2882
10. Kaur Kang, G., Gao, J.Z., Chiao, S., Lu, S., Xie, G.: Air quality prediction: big data and machine learning approaches. Int. J. Sci. Technol. Res. (2020)
11. Ambika, G., Bhanu Pratap Singh, N., Dishi, B.S.: Tiwari air quality index prediction using linear regression. Int. J. Recent Technol. Eng. (IJRTE) **8**(2) (July 2019). ISSN: 2277-3878
12. Mahesh Babu, K., Rene Beulah, J.: Air quality prediction based on supervised machine learning methods. Int. J. Innovative Technol. Exploring Eng. (IJITEE) **8**(9S4) (July 2019). ISSN: 2278-3075
13. Madan, T., Sagar, S., Virmani, D.: Air quality prediction using machine learning algorithms –a review. In: 2020 2nd International Conference on Advances in Computing, Communication Control and Networking (ICACCCN), pp. 140–145 (2020). https://doi.org/10.1109/ICACCCN51052.2020.9362912

14. Sujatha, G., Vasantha Kumari, P., Vanathi, S.: Air quality indexing system –a review. AICTE Sponsored Two Days Online National Level E-Conference On Machine Learning as a Service for Industries MLSI 2020 4th–5th September (2020). ISBN: 978-93-5416-737-9

15. https://www.kaggle.com/rohanrao/noise-monitoring-data-in-indiaauthor, F.: Article title. Journal 2(5), 99–110 (2016)

Mura Defect Detection in Flat Panel Display Using B-Spline Approximation

Kishore Kunal[1(✉)], Pawan Kumar Upadhyay[2], M. Ramasubramaniam[1],
and M. J. Xavier[1]

[1] Centre for Technology and Innovation, LIBA Chennai, Chennai, India
kishore.sona@gmail.com
[2] Jaypee Institute of Information Technology, Noida, India

Abstract. Flat panel display (FDP) devices continue to grow at rapid
rate and quite popular as a promising technology and evolve various
investment opportunity. In this paper, a machine vision approach has
been proposed for automatic inspection of mura defects in film or glass of
flat panel display device. The proposed method is based on a Mura filter
using B-spline global approximation method. Experimental result shows
that the detection of mura defect has been performed on images data
with high speed computational techniques and obtained robust results.

Keywords: B-spline · Machine learning · Machine vision · Flat panel
display

1 Introduction

Thin film transistor liquid crystal display (TFT-LCD) devices have become a
major innovative technology for FPD. The governing display unit have become
equally important in recent years, due to their full color display capabilities, low
power consumption and light weight. Flat panel display can be used as monitor,
for notebook and PC's and as viewfinders for hand held devices such as cellular
phone and PDA's. Due to its high demand in market, the quality of the display
becomes a more critical issue for manufacturers. In order to ensure the display
quality and improve the yield of FPD panels, the inspection of defects in FPD
panels becomes a tedious task in manufacturing. Human visual inspection which
is still used by the most manufacturers has a number of drawbacks including the
limitations of human sensibility, inconsistent detection due to human subjectiv-
ity and high cost. However, manual inspection is a time consuming and tiresome
assignment. It highly dependent on the experience of human inspectors. Automatic inspection using machine vision techniques can overcome many of these
disadvantages and offer manufacturers an opportunity to significantly improve
the quality and reduce cost. One class of defects includes Mura (*Japanese syn-
onms for Artifects*) has been widely adopted by the display industry to describe

Supported by organization LIBA Chennai.

almost all capricious stain variation defects in FPD [2]. Mura defects are caused by process faults usually related to cell assembly, which affects the transmission of light through display [4]. It is extremely difficult for FPD manufacturers to make accurate detection and classification due to its cyclical nature, randomness and often low contrast [1]. They are caused by a variety of physical factor such as non-uniformly distributed liquid crystal material and foreign particles within the liquid crystal. Depending on the shapes and sizes, mura defects may be classified into spot mura, line mura and area mura defect which is depicted in Fig. 1. As compared to spot mura and line mura, area mura is relatively difficult to identified due to its low contrast and no particular pattern of shape [5].

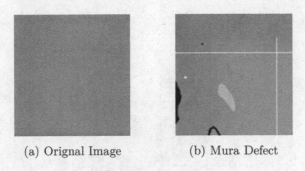

(a) Orignal Image   (b) Mura Defect

Fig. 1. Example of original image and mura defect

In this paper we thus present the technique focused on all type of mura by describing an automatic detection method that reliably detects and quantifies FPD mura defects by remarkably considering with our approach in Sect. 3 and motivation behind the approach is significantly discuss in Sect. 2 of this paper.

2   B-Spline Approximation

Splines are piecewise polynomials with pieces that are smoothly connected together [6]. The joining points of the polynomials are called Knots [7]. Recently, B-spline approximation technique has been proposed for image reconstruction to detect the mura defects [3]. In this section we elaborate on that technique in terms of scattered data interpolation and present details of algorithm. Given a set of scattered points,

$$P = \{p_i\}_{i=1}^n, p_i = (x_i, y_i, z_i) \in R^3 \tag{1}$$

and let,

$$\Omega = \{(x, y) \mid 0 \le x \le m_x 0 \le y \le m_y\} \tag{2}$$

be a rectangular domain in the x- y plane such that (x_i, y_i) is a point in Ω. Let ϕ be a control lattice overlaid on a domain Ω. The control lattice ϕ is a uniform tensor product grid over Ω. To approximate scattered data points P, we formulate initial approximation function as a uniform bi-cubic B-spline function f, which is defined by a control lattice ϕ. Let the initial number of control points on the lattice as $n_x = m_x / h_x$ in x axis and as $n_y = m_y / h_y$ in y axis. The knot intervals are uniform defined as in x axis and in y-axis. So for uniform cubic B-spline case, degree d = 3 and the set of knot vectors are defined as below (Fig. 2):

$$\tau_x = \{-dh_x, \ldots, 0, h_x, \ldots n_x h_x, \ldots (n_x + d) h_x\}$$
$$\tau_y = \{-dh_y, \ldots, 0, h_y, \ldots n_y h_y, \ldots (n_y + d) h_y\} \tag{3}$$

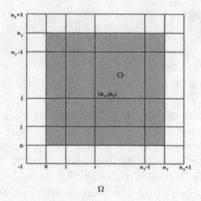

Ω

Fig. 2. The configuration of control lattice Ω

Let c_{ij} be the value of the ij-th control point on lattice, located at position of the grid defined as, for i = 1, 0, 1, ..., (n_x+1) and j = −1, 0, 1, ..., (n_y+1). The approximation function f defined in terms of these control points at position (x, y) denoted as Ω is computed as:

$$\tau_x = \{-dh_x, \ldots, 0, h_x, \ldots n_x h_x, \ldots (n_x + d) h_x\}$$
$$\tau_y = \{-dh_y, \ldots, 0, h_y, \ldots n_y h_y, \ldots (n_y + d) h_y\} \tag{4}$$

where $B_{i,d}$, are uniform cubic B-spline basis functions of d = 3 and knot vector for cubic B-spline basis are below:

$$\{(i-2)h_x, (i-1)h_x, ih_x, (i+1)h_x, (i+2)h_x\}$$
$$\{(j-2)h_x, (j-1)h_x, jh_x, (j+1)h_x, (j+2)h_x\} \tag{5}$$

To optimize this process, B-spline refinement is used to reduce the sum of these functions into one equivalent B-spline function. The method explored in this paper deals with uniform data of an input image as a data. The method govern with the following steps as (1) consider an input image block of size (256 * 256) which is show in Fig. 3(a) in 3D having x and y as indices and z as color gray values of the image. In continuation of this, (2) select a set of 7×7 neighboring pixels as an control points and implementing cubic B-spline approximation technique. (3) Repeat the process, till complete block covered and the obtained result are profoundly depicted in Fig. 3, as shown describe below:

(a) 256x256 bitmap image   (b) Cubic B-spline approximation

Fig. 3. 7×7 control points over (a) Input images to perform approximation using cubic B-Spline and generate (b) Approximated image

The above steps for defect detection is the global approximation technique. In the above approach, there is huge significance of pre-calculate weight, as obtained coefficients from 4×4, 7×7 control points. These pre-calculated weights remarkably increases the computational speed and reduces the overall cost of the system. The related results are shown in the Fig. 4. In addition to this, the image having 7×7 controls points has more degree of freedom, so it gives more accuracy than that of 4×4 control points. Furthermore, the 4×4 control points have some advantages like they generates smooth surface for FPD on computation.

(a) 4x4 control points   (b) 7x7 control points

Fig. 4. Image control points

3 Proposed Approach: Mura Detection Algorithm

The algorithm explored in this paper take B-spline approximation from original image, and calculates difference between original and approximation image as input and produces approximated refine image as an output. We first find the non negative pixel value in 3×3 mask and then take the mean of that gray level values. The algorithm implemented on 3072×6138 image size which takes 9.118 ms to execute. The following outcome will come after applying mura detection algorithm which is shown in Fig. 5. The algorithm can be summarized as follows:

Algorithm 1. Mura Detection Algorithm

I: Input Image
O: Output Image

       1. $I_{i,j} : \{I(i-1, j-1), I(i-1, j), \ldots I(i+1, j+1)\}$
       2. $P_{i,j} :$ Non negative values of $I_{i,j}$
       3. $P1_{i,j} :$ Number of $P_{i,j}$
       4. $mP_{i,j} :$ Mean of $P_{i,j}$
       5. if $(\#P1_{i,j} > 7)\, O(1, j) = mP_{i,j}$
       6. else iff $\#P1_{i,j} < 2)\, O(i, j) = m\left(I_{i,j} - P_{i,j}\right)$
       else $O(i, j) = mI_{i,j}$

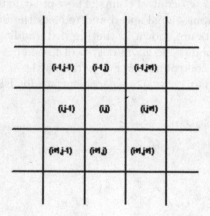

Fig. 5. Depict the pixels adjacency for identifying the Mura defects

(a) 4x4 control points   (b) 7x7 control points

Fig. 6. Image control points

4   Experimental Results

To demonstrate the accuracy of detection by the proposed algorithm, we performed experiments on standard system of HP workstation xw4200, P4 3.4 GHz, 1 GB RAM. First, our proposed image with Mura detection filter is compared with the number of varying control points. In the given table, we compared the initial image size of 3072×6138 and taking the block size of 1024×1024 by varying the different types of control points. We compute time of all image block using control points and applying the cubic B-spline approximation technique in terms of execution time. Here is the standard result which is illustrated in Table 1 as depicted below:

Table 1. Comparisons of block images and their execution time using different control points

Block size	Number of control points	Seconds
256	7	0.672
	5	0.468
	4	0.375
128	7	0.672
	5	0.453
	4	0.375
64	7	0.687
	5	0.454
	4	0.375

According to table, we see that the computation time of image block having 4×4 control points is significantly better than that of other. So, we choose 256×256 bitmap image block having 4×4 control points to detect the defect. The final result shown in figure that demonstrates the time efficient approach using 256×256 block images to detect the mura defect.

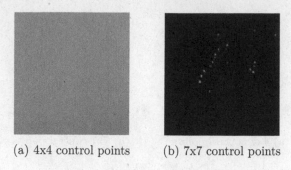

(a) 4x4 control points   (b) 7x7 control points

Fig. 7. Image control points

5   Conclusion

In this paper, a B-spline approximation based technique have been proposed for uniform data. The algorithm is fast and able to detects mura defect through a set of control points considerably know as mura filter. The experimental result reveals that the proposed approach is not only robust but computationally efficient. In addition to this, It also able to detect all types of effected region. The naive approach helps to perform mura level quantification more efficiently and able to generate consistent quality of visuals in Flat panel display.

References

1. Chuang, Y.C., Fan, S.K.S.: Automatic detection of region-MURA defects in TFT-LCD based on regression diagnostics (2009)
2. Fan, S.K., Chuang, Y.C.: Automatic detection of Mura defect in TFT-LCD based on regression diagnostics. Pattern Recogn. Lett. **31**, 2397–2404 (2010). https://doi.org/10.1016/j.patrec.2010.07.013
3. Jiang, Z.: A new approximation method with high order accuracy. Math. Comput. Appl. **22**(1) (2017). https://doi.org/10.3390/mca22010011, https://www.mdpi.com/2297-8747/22/1/11
4. Kunal, K., Prasad, A., Xavier, M., Arun, J.: Mura defect detection in LCD. Acad. Mark. Stud. J. **25**(5), 1–10 (2021). https://doi.org/10.1016/j.patrec.2010.07.015
5. Lee, B.G., Lee, J.J., Yoo, J.: An efficient scattered data approximation using multilevel B-splines based on quasi-interpolants. **2**(5), 110–117 (2005). https://doi.org/10.1109/3DIM.2005.18
6. Lee, S., Wolberg, G., Shin, S.: Scattered data interpolation with multilevel b-splines. IEEE Trans. Visual Comput. Graphics **3**(3), 228–244 (1997). https://doi.org/10.1109/2945.620490
7. Unser, M.: Splines a perfect fit for signal and image processing. IEEE Signal Process. Mag. **16**(6), 22–38 (1999). https://doi.org/10.1109/79.799930

An Efficient Sample Steering Strategy for Correlation Filter Tracking

S. M. Jainul Rinosha$^{(\boxtimes)}$ and M. Gethsiyal Augasta

Research Department of Computer Science, Kamaraj College (Affiliated to Manonmaniam Sundaranar University, Tirunelveli), Thoothukudi, Tamilnadu, India
sm.jainulrinosha@gmail.com

Abstract. Nowadays, many researchers utilize the correlation filter based trackers to address the severe appearance and motion changes in video sequences. A correlation filter (CF) can be evaluated based on the relationship between the object ground truth region (local) and the steering samples region (semi-local), with spatial and visual selective attention in parallel to human visual perception. In the proposed work, initially, an efficient correlation filter is estimated with the circulant sample matrices and the steering strategy. To effectively remove the boundary effect of the CF, the Histogram of Gradients (HoGs) are extracted for positive and negative samples which in turn referred the semi-local domains. Finally, the linear combination of the two domain models is used to determine the location of the object. The objective of the proposed visual tracking method (OT-Steer) is to apparently minimise the distractions and to effectively remove the boundary using the combination of object tracking and steered approach. The proposed OT-Steer model has been evaluated with various tracking benchmark datasets and results are compared. The results depict that the OT-Steer plays a vital role in effectively dealing with distractions during the tracking process with the low computational cost for efficient visual tracking.

Keywords: Visual tracking · Circulant matrix · Correlation filter · Histogram of gradient · Object tracking · Steering approach

1 Introduction

Advanced research in Visual Object Tracking is an emerging field for computer vision problems to track a general target in an abandoned input frame sequence. Visual tracking is a challenging task for the real-world application because an object, in uncontrollable recording conditions is usually affected by huge illumination variations, size differences, background clutters, and dense occlusions to identify a particular object in all the frames of videos and place it only in the initial frame. The application of visual tracking is extensively used in the fields such as surveillance, traffic monitoring, disaster response, person tracking, human-machine interaction, unmanned vehicles, defense, and crowd behaviour investigation. There are different methodologies opted for tracking an object, they are namely; Discriminative Trackers, Generative Trackers, Correlation Filter Based

© Springer Nature Switzerland AG 2022
R. Chbeir et al. (Eds.): MIKE 2021, LNAI 13119, pp. 109–119, 2022.
https://doi.org/10.1007/978-3-031-21517-9_11

Trackers and Combined Trackers [1]. Enormous numbers of tracking algorithms have been implemented in Correlation Filter (CF) based tracking domain. Usually, the weights of the CF learnt from the given samples are utilized for further tracking process. In order to learn the weights effectively, lot of efforts have been placed in recent days. Still, identifying the improved filter adoption process is an essential aim for any correlation filter based tracking principles.

2 Related Works

Nowadays, researchers have proposed various correlation filter based visual tracking algorithms [2–9]. Li et al. [10] have presented the target-aware correlation filters (TACF) and an optimization strategy based on preconditioned conjugate gradient method. Zhang et al. [11] have proposed the spatially Regularized Discriminative Correlation Filters (SRDCF) which define multiple correlation filters to extract features and build part-based trackers with the combination of the cyclic object shift and penalized filters. Sun et al. [12] have presented a novel ROI Pooled Correlation Filter (RPCF) algorithm by using mathematical derivations. It learns from the weights of the filter and makes the ROI-based pooling. Dai et al. [13] have presented an adaptive spatially-regularized correlation filters model based on the alternating direction method of the multiplier, and trackers with two kinds of CF models to find the location and scale. An adaptive model and tracking framework have been proposed by Wang et al. [14] which develop the relationship between the target and its spatiotemporal context model in a hierarchical way. Zheng et al. [15] have proposed a multi-task deep dual correlation filters which perform against the state-of-the-art method. A parallel correlation filter framework for visual object tracking was introduced by Yang et al. [16], in which initially two parallel CFs are constructed to track the appearance changes and translation of the target. Then the weighted response maps are merged and finally, the correlation output is distributed. Li et al. [17] have proposed a multi-branch model and adaptive selection strategy to tolerate sequential changes of the object served in various conditions. It includes both the foreground and background evidence to learn the background suppression. MACF framework for online visual tracking was proposed by Zhang et al. [18] using the DCF-based method, to detect the position and scale of the target. A hybrid tracker was proposed by Xia et al. [19] with the combination of the deep feature method and correlation filter to make multi-layer feature fusion. Sun et al. [20] have proposed a robust tracking method that uses the multiple tracking dataset to display the accuracy of the tracker by comparing it to the state-of-the-art based on the deep learning features. Li et al. [21] have designed an ADCF tracker for the translation of the target with the time complexity O(NlogN) and space complexity O(N), an algorithm has been designed for multiplying two-level block Toeplitz matrix.

A location-aware and regularization-adaptive correlation filter (LRCF) was designed by Ding et al. [22], which include a LRCF-S and LRCF-SA trackers to estimate the location and filter the problems during the training process. Zuo et al. [9] have derived an SVM model with circulant matrix expression for visual tracking and also extended the SCF-based tracking method to improve the tracking performance. An iteration of the Alternating Direction Method of Multipliers (ADMM) was proposed by Liang et al.

[23] that describe a Spatio-temporal adaptive and channel selective correlation filter. Initially, the targeted features are selected based on the Taylor expansion. Then, the filter learning problem is reformulated from the ridge regression for robust tracking. An SVM-based tracking method was proposed by Su et al. [24] using the classifier weights and kernel weights, and it was computed by fast Fourier transform. When analysing the state of the strategies which are available in the literature, the correlation filter based tracking approaches struggle with the distraction while the target object occluded with some featured object. Zhang et al. [25] introduced a novel distractor-resilient metric regularization along with CF to process the local and semi-local domains. The method push distractors into negative space, controls the Boolean map representation and tracks the non-rectangular objects. Augasta et al. [26] developed a visual tracking method known as OT-CS to reduce the distractors and to remove the boundary using the linear combination of local and semi local domains for object tracking.

The main objective of this research is to improve the correlation filter based visual tracking with the steering strategies. To overcome the disadvantages of the tracking process with the aid of correlation filter methods, a methodology called OT-Steer has been introduced with the steering strategies that calculate the score to select the highly correlated steering samples, resulting in enhanced efficiency and better results. In OT-Steer methodology, the effective tracking architecture has been achieved by using circulant sample generation, followed by steering the samples, adhering to a feature map-based correlation response, and then tracking the object. This paper is organized as follows: Sect. 2 explains the methodology of the proposed work, Sect. 3 presents the experimental results on experimental datasets and Sect. 4 compares the performance of the proposed method with other tracking methods.

3   Methodology

In the context of visual tracking, several techniques have been employed and modified to calculate the response map for tracking. Distractors can lead to failure when computing the response map, whereas an anti-distractor approach can lead to higher monitoring performance. In the proposed method, an efficient correlation filter is estimated with the circulant matrix and the steering strategy. It removes the background effect of the correlation filter and learns the HOG for both the ground truth region and steering sample region. The proposed method is named as Object Tracking using Steering Strategy (OT-Steer). The pictorial representation of the proposed architecture is presented in Fig. 1.

3.1   Circulant Sample Generation

In OT-Steer, object tracking is achieved with the help of both the correlation filters and matrices of the circulant samples. In a circulant matrix, elements have a fixed value along the main diagonals and sub-diagonals, and it is organized by the cyclically shifted versions of a length of a particular n-list. Circulant matrices are most effective and useful in digital image processing. While exploring the structure of the circulant matrix for high efficiency, many negative samples are employed to enhance the ability to track by detector scheme. Here, the target of the image is fed as an input, and from that input image, the circulant matrices are derived to hold the numerous image samples. An appearance of a target object using a correlation filter 'c' trained on an image patch p of R × S pixels, where all the circular samples of p, r, s, (r, s) ∈ {0, 1, ..., R − 1} × {0, 1, ..., S − 1}, are generated as training samples with a Gaussian function label $g_{r, s}$. The goal is to find the optimal weights c,

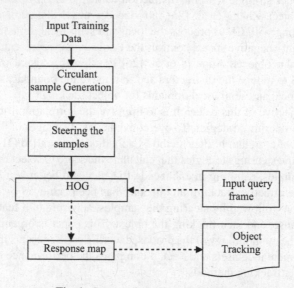

Fig. 1. Proposed tracking architecture

$$c = arg_c min \sum_{r,s} \left| (\phi(x_{r,s}), c) - g_{r,s} \right|^2 + \lambda ||c||^2 \tag{1}$$

where φ denotes the mapping to a kernel space and λ is a regularization parameter. This objective function is minimized as $c = \sum_{r,s} \propto (r, s)\phi(p_{r,s})$ by using the Fast Fourier Transformation (FFT) to compute the correlation, and the coefficient α is calculated as

$$\alpha = F^{-1}\left(\frac{F(g)}{F((\phi(p), \phi(p))) + \lambda} \right) \tag{2}$$

where g = {g(r, s)}, F denotes the Fourier Transform and F^{-1} denotes the Fourier transform inverse, respectively.

3.2 Steering the Samples

After generating the numerous image samples by using the circulant matrices along with its coefficients $\propto_i$, i∈{1, 2, ..., n} where n is number of circulant samples, the input samples are steered with the help of correlation, overlap ratio and ground truth. The bounding boxes for all steering samples will be drawn prior to commence the steering process. Let γ_i be the overlap ratio between object region and steering samples. The overlap ratio γ_i is defined as in Eq. (3).

$$\gamma_i = \frac{Ao_i}{Au_i} \tag{3}$$

where, Ao_i denotes the area of overlap between the ground truth and i^{th} sample region. Au_i denotes the area encapsulated by both the ground truth and i^{th} sample region. The score s_i for each sample is calculated using the Eq. (4),

$$s_i = \frac{(\alpha_i + \gamma_i)}{2} \tag{4}$$

After calculating the score for all samples, selection of the steered sample images is done with the condition that $s_i > k$ where k is the threshold value lies between 0 to 1. The satisfied samples are considered to be steered samples. These steered samples are selected to learn the histogram of gradients features.

3.3   Feature Map Based Correlation Response

The HOG features are used to learn the CF, based on positive and negative samples or ground truth region (local domain) along with steering samples (semi-local domain). The response map is generated at the end of the feature-based correlation response by using the HOG features. HOG is a feature based extraction algorithm that converts the image with the fixed size into a feature vector of a particular size. HOG Descriptor calculates the gradient function of an image, divides into 8×8 cells, calculates the histogram of the gradient of the divided cells, multiplies the image intensities to produce the block normalization, calculates the feature vector of an entire image, and finally generates the correlation response map using the local and semi local domain.

3.4   Object Tracking Process

This section describes the object tracking process of the proposed method OT-Steer. Initially, the positive samples from the initial object region are extracted and the circulant sample matrices are generated according to the procedure discussed in Sect. 3.1. Then, some of the prominent i.e., steered samples are selected by using the steering process in which the score was calculated to select the particular samples with a better score. For obtaining score, the bounding boxes are drawn earlier for all steering samples to begin the steering process. Then, the overlapping ratio among the object region and steering samples are obtained. Finally, scores are calculated (using Eq. 4) for each samples and validates the score using condition to select the steered samples. As a next step, the steered samples are passed inside the HOG features. Here, it extracts the set of HOG for both the positive and negative domain. And, it generates the response map based on the learned weight. Finally, the proposed method tracks the object by using the response map subsequently for the next time frame.

4   Experimental Results and Analysis

4.1   Experimental Setup

All experiments were implemented in MATLAB on a PC i3 3.6 GHz CPU and 8 GB RAM. The parameters like regularization factor λ and the learning rate is fixed as 0.025 0.0625 respectively and the number of padding pixels is considered as 2. The selection threshold k is considered as greater than or equal to 0.6. Moreover, the parameters are fixed as same for all experimental datasets.

4.2   Datasets

The proposed method is evaluated on a benchmark dataset namely Dog, Biker, Blur Car, Basket Ball, Face and Bird [26, 27] and OTB 2015 [28], VOT 2016 [29], and MOT 20 [30] which have video sequences of data already labeled with the attributes based on various features in visual tracking. OTB 2015 [28] dataset has 100 video sequences to evaluate visual tracking, while VOT 2016 [29] dataset has 60 video clippings and 21,646 corresponding ground truth maps with pixel-wise annotation of the objects. MOT 20 [30] dataset contains 8 video sequences for evaluating the visual tracking. The subset of this benchmark is taken into consideration for experimental processes with different combinations and attributes.

4.3   Results

In order to evaluate the proposed tracker OT-Steer, the Total Elapse Time for tracking speed, the accuracy, center location error and average error plot are estimated. Accuracy is computed by finding the whole difference between the ground truth and the desired frame. The experimental results of OT-Steer on nine benchmark datasets namely Dog, Biker, Blur Car, Basket Ball, Face, Bird, OTB 2015, VOT 2016, and MOT 20 are shown in Table 1. The results depicts that the proposed tracker OT-Steer achieves 100% accuracy for the face dataset and enhanced accuracy for other experimental datasets in less tracking time.

The output sequences of the proposed method OT-Steer on two datasets namely Face and Dog are depicted in Fig. 2. The output sequences of the proposed method OT-Steer on two datasets namely Face and Dog are depicted in Fig. 2. Here the generated samples using circulant transition matrix, steered samples using OT-Steer and finally the tracked object are presented sequentially in figures a, b and c.

Table 1. Experimental results of OT-steer

Input	Elapsed time (s)	Accuracy	Center location error	Average error
Dog	13.653	0.950	0.040	0.050
Biker	15.355	0.863	0.075	0.078
Blur car	15.856	0.823	0.075	0.314
Basketball	15.927	0.852	0.768	0.708
Face	13.085	1.000	0	0
Bird	15.314	0.550	0.450	0.450
OTB 2015	14.452	0.852	0.047	0.053
VOT2016	13.523	0.999	0.075	0.068
MOT 20	13.24	0.894	0.074	0.041

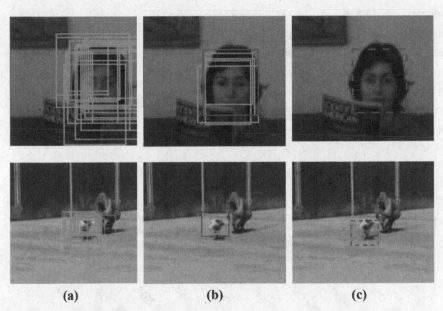

(a)   (b)   (c)

Fig. 2. The output sequences of OT-Steer on two datasets (i. Face and ii. Dog) (a) generated samples using circulant transition matrix (b) steered samples (c) Tracked object.

5   Performance Analysis

In this section, the performance of proposed tracker OT-Steer is compared with the similar state of art trackers OT-BMR [29] and OT-CS [30]. Table 2 compares OT-Steer in terms of Total Elapsed Time and Accuracy with OT-CS and OT-BMR. The comparative results in Table 2 show that the OT-Steer performs outstandingly than other related trackers OT-CS and OT-BMR. Considering the tracking speed, OT-Steer outperforms the other trackers in study by consuming less tracking time.

In terms of accuracy, OT-Steer achieves higher accuracy for seven experimental datasets out of 9. The proposed method OT-Steer has an overall average measure as 0.865. At the same time, the method OT-CS and the method OT-BMR has an overall average measure value as 0.817 and 0.708 respectively. While comparing the overall average measures, the proposed OT-Steer method obtains better accuracy than other methods in comparison. Figure 3 shows the comparison results of OT-Steer in terms of accuracy.

Table 2. Comparative analysis of OT-steer in terms of elapsed time and accuracy

Input	Elapsed time			Accuracy		
	OT-Steer	OT-CS	OT-BMR	OT-Steer	OT-CS	OT-BMR
Dog	**13.653**	17.704	24.146	**0.950**	0.947	0.893
Biker	**15.355**	18.303	24.099	0.863	**0.920**	0.907
Blur car	**15.856**	18.337	23.773	**0.823**	0.653	0.507
Basketball	**15.927**	18.588	23.493	**0.852**	0.792	0.754
Face	**13.085**	17.308	22.908	**1.000**	0.999	0.999
Bird	**15.314**	18.743	24.899	0.550	**0.587**	0.413
OTB 2015	**14.452**	15.74	19.154	**0.852**	0.793	0.623
VOT 2016	**13.523**	14.672	17.192	**0.999**	0.834	0.634
MOT 20	**13.24**	15.653	17.342	**0.894**	0.832	0.643

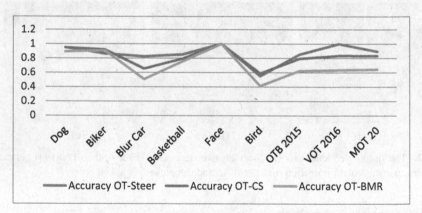

Fig. 3. Comparative analysis of OT-steer in terms of accuracy

In addition, the proposed method OT-Steer is analysed in terms of center location error and Average error. Table 3 compares the results of Center Location Error and Average Error for the methods OT-Steer, OT-CS and OT-BMR. Results depict that the OT-Steer performs visual tracking with minimum errors while comparing with other trackers.

Table 3. Comparison of OT- Steer in terms of Center Location Error and Average Error

Input	Center location error			Average error		
	OT-Steer	OT-CS	OT-BMR	OT-Steer	OT-CS	OT-BMR
Dog	**0.040**	0.040	0.080	**0.050**	0.053	0.107
Biker	**0.075**	0.082	0.093	**0.078**	0.080	0.093
Blur Car	**0.075**	0.081	0.087	**0.314**	0.347	0.493
Basketball	**0.768**	0.773	0.773	**0.708**	0.773	0.773
Face	**0**	0	0	**0**	0.001	0.001
Bird	**0.450**	0.519	0.524	**0.450**	0.413	0.587
OTB 2015	**0.047**	0.04	0.08	**0.053**	0.069	0.107
VOT 2016	**0.075**	0.082	0.0933	**0.068**	0.08	0.093
MOT 20	**0.075**	0.081	0.087	0.041	**0.035**	0.049

Moreover, conceptually OT-BMR [29] fully integrates local domain features into the correlation filter validation process, while OT-CS [30] eliminates distractions using graceful selective shifted patches. As a result, the boundary effect is resisted and the tracking accuracy is improved. Beyond that, OT-Steer not only precisely eliminates the distractors and also it selects the samples dynamically based on the relationship between the regions. So, the proposed method foots forth the trackers efficiently in terms of different performance measures.

6 Conclusion

Visual tracking is an essential building block of many advanced applications in the areas such as video surveillance and human-computer interaction. In this research, the steering mechanism with circulant matrices is considered as key aspect to enhance the correlation filter based visual tracking. The proposed OT-Steer works by extracting the positive samples, generating the circulant sample matrices, calculating the scores and selecting the steering samples with a better score. Then extracts the HOG features of the steering samples and generates the response map according to the learned weight. Finally tracks the object for the next instance. The experiments have been extensively conducted on benchmark datasets and the results analysis show that the proposed OT-Steer tracker has highly contributed to track the targets in a better manner. Further work has been planned to automate the tracking by employing the OT-Steer methodology with deep learning.

References

1. Jainul Rinosha, S.M., Gethsiyal Augasta, M.: Review of recent advances in visual tracking techniques. Multimed. Tools Appl. **80**, 24185–24203 (2021)

2. Li, F., Tian, C., Zuo, W., Zhang, L., Yang, M.H.: Learning spatial-temporal regularized correlation filters for visual tracking. In: Proceedings of the IEEE Conference on Computer Vision and Pattern Recognition, pp. 4904–4913 (2018)
3. Wang, Q., Gao, J., Xing, J., Zhang, M., Hu, W.: Dcfnet: Discriminant correlation filters network for visual tracking. arXiv preprint arXiv:1704.04057 (2017)
4. Wang, N., Zhou, W., Tian, Q., Hong, R., Wang, M., Li, H.: Multi-cue correlation filters for robust visual tracking. In: Proceedings of the IEEE Conference on Computer Vision and Pattern Recognition, pp. 4844–4853 (2018)
5. Bai, S., He, Z., Xu, T.B.: 1811. Multi-hierarchical independent correlation filters for visual tracking, arXiv preprint arXiv:1811.10302 (2018)
6. Zhang, M., et al.: Visual tracking via spatially aligned correlation filters network. In: Proceedings of the European Conference on Computer Vision (ECCV), pp. 469-485 (2018)
7. Chen, Z., Guo, Q., Wan, L., Feng, W.: Background-suppressed correlation filters for visual tracking. In: 2018 IEEE International Conference on Multimedia and Expo (ICME), pp. 1–6. IEEE (2018)
8. Liu, R., Chen, Q., Yao, Y., Fan, X., Luo, Z.: Location-aware and regularization-adaptive correlation filters for robust visual tracking. IEEE Trans. Neural Netw. Learn. Syst. 32(6), 2430–2442 (2020)
9. Zuo, W., Wu, X., Lin, L., Zhang, L., Yang, M.H.: Learning support correlation filters for visual tracking. IEEE Trans. Pattern Anal. Mach. Intell. 41(5), 1158–1172 (2018)
10. Li, D., Wen, G., Kuai, Y., Xiao, J., Porikli, F.: Learning target-aware correlation filters for visual tracking. J. Vis. Commun. Image Represent. 58, 149–159 (2019)
11. Zhang, D., et al.: Part-based visual tracking with spatially regularized correlation filters. Vis. Comput. 36(3), 509–527 (2019). https://doi.org/10.1007/s00371-019-01634-5
12. Sun, Y., Sun, C., Wang, D., He, Y., Lu, H.: Roi pooled correlation filters for visual tracking. In: Proceedings of the IEEE/CVF Conference on Computer Vision and Pattern Recognition, pp. 5783–5791 (2019)
13. Dai, K., Wang, D., Lu, H., Sun, C., Li, J.: Visual tracking via adaptive spatially-regularized correlation filters. In: Proceedings of the IEEE/CVF Conference on Computer Vision and Pattern Recognition, pp. 4670–4679 (2019)
14. Wang, W., Zhang, K., Lv, M., Wang, J.: Hierarchical spatiotemporal context-aware correlation filters for visual tracking. IEEE Trans. Cybern. (2020). https://doi.org/10.1109/TCYB.2020.2964757
15. Zheng, Y., Liu, X., Cheng, X., Zhang, K., Wu, Y., Chen, S.: Multi-task deep dual correlation filters for visual tracking. IEEE Trans. Image Process. 29, 9614–9626 (2020)
16. Yang, Y., Zhang, Y., Li, D., Wang, Z.: Parallel correlation filters for real-time visual tracking. Sensors 19(10), 2362 (2019)
17. Li, X., Huang, L., Wei, Z., Nie, J., Chen, Z.: Adaptive multi-branch correlation filters for robust visual tracking. Neural Comput. Appl. 33(7), 2889–2904 (2020). https://doi.org/10.1007/s00521-020-05126-9
18. Zhang, Y., Yang, Y., Zhou, W., Shi, L., Li, D.: Motion-aware correlation filters for online visual tracking. Sensors 18(11), 3937 (2018)
19. Xia, H., Zhang, Y., Yang, M., Zhao, Y.: Visual tracking via deep feature fusion and correlation filters. Sensors 20(12), 3370 (2020)
20. Sun, Z., Wang, Y., Laganière, R.: Hard negative mining for correlation filters in visual tracking. Mach. Vis. Appl. 30(3), 487–506 (2019). https://doi.org/10.1007/s00138-019-01004-0
21. Li, S.-W., Jiang, Q., Zhao, Q.-J., Lu, L., Feng, Z.-L.: Asymmetric discriminative correlation filters for visual tracking. Front. Inform. Technol. Electron. Eng. 21(10), 1467–1484 (2020). https://doi.org/10.1631/FITEE.1900507

22. Ding, G., Chen, W., Zhao, S., Han, J., Liu, Q.: Real-time scalable visual tracking via quadrangle kernelized correlation filters. IEEE Trans. Intell. Transp. Syst. **19**(1), 140–150 (2017)
23. Liang, Y., Liu, Y., Yan, Y., Zhang, L., Wang, H.: Robust visual tracking via spatio-temporal adaptive and channel selective correlation filters. Pattern Recogn. **112**, 107738 (2021)
24. Su, Z., Li, J., Chang, J., Du, B., Xiao, Y.: Real-time visual tracking using complementary kernel support correlation filters. Front. Comp. Sci. **14**(2), 417–429 (2019). https://doi.org/10.1007/s11704-018-8116-1
25. Gethsiyal Augasta, M., Jainul Rinosha, S.M.: Correlation filter based visual tracking with circular shift on local and semi-local domains. Int. J. Future Gener. Commun. Netw. **13**(2), 941–948 (2020)
26. http://cvlab.hanyang.ac.kr/tracker_benchmark/datasets.html
27. Wu, Y., Lim, J., Yang, M.-H.: Online object tracking: a benchmark. In: Proceedings of the IEEE Conference on Computer Vision and Pattern Recognition, pp. 2411–2418 (2013)
28. Huang, Y., Zhao, Z., Wu, B., Mei, Z., Cui, Z., Gao, G.: Visual object tracking with discriminative correlation filtering and hybrid color feature. Multimedia Tools Appl. **78**(24), 34725–34744 (2019). https://doi.org/10.1007/s11042-019-07901-w
29. Kristan, M., et al.: The visual object tracking VOT2016 challenge results. In: Hua, G., Jégou, H. (eds.) ECCV 2016. LNCS, vol. 9914, pp. 777–823. Springer, Cham (2016). https://doi.org/10.1007/978-3-319-48881-3_54
30. Dendorfer, P., et al.: MOT20: a benchmark for multi object tracking in crowded scenes. arXiv: 2003.09003 [cs] (2020)

A Comparative Study on Machine Learning Based Classifier Model for Wheat Seed Classification

G. Sujatha[1] and K. Sankareswari[2(✉)]

[1] Department of Computer Science, Sri Meenakshi Govt. Arts College for Women (A), Madurai, India
[2] Department of Computer Science, The American College, Madurai, India
sankari.kaverimanian@gmail.com

Abstract. Seed classification is a process of categorizing different varieties of seeds into different classes on the basis of their morphological features. Seed identification is further complicated due to common object recognition constraints such as light, pose and orientation. Wheat has always been one of the globally common consumed foods in India. A large number of wheat varieties have been cultivated, exported and imported all around the world. Enormous studies have been done on identifying crop diseases and classifying the crop types. In the present work, wheat seed classification is performed to distinguish the three different Indian wheat varieties by their collected morphological features and applied machine learning models to develop wheat variety classification system. The seed features used here are length of kernel, compactness, asymmetry coefficient, width of kernel, length of kernel groove, area and perimeter. The present work carried out with different classifiers such as Decision Tree, Random Forest, Neural Net, Nearest Neighbors, Gaussian Process, AdaBoost, Naive Bayes, Support Vector Machine(SVM) Linear, SVM RBF(SVM with the Radial Basis Function) and SVM Sigmoid with 2 K-fold cross validation. Also obtained the results using 5 fold and 10-fold Cross Validation.

Keywords: Agriculture · Seed classification · SVM · Neural net · Gaussian · K-fold cross validation · Random forest · Decision tree · Classifiers

1 Introduction

As essential food of thousands of millions of people in the world, totally, wheat's production amounted to 772 million tonnes each year [8], while its supply was just 179.26 g per day per capita [13] in the world. Agriculture is a most important sector in Indian economy. As agriculture struggles to support the rapidly growing global population, seed quality is the most important factor to crop production and quality of food [9]. Seeds with good quality plays a crucial role in the crop yield improvement. Seeds with low quality will leads to loss to the farmers in terms of planting, breeding and crop quality [1, 2, 4]. In Modern seed industry, accidental mixing between different variety of wheat seeds during storage, transportation and production which result in poor quality and yield. Therefore, rapid approach has to be needed to identify and classify different varieties of wheat seeds. Normally, classification of different varieties of wheat seeds are determined with respect to varietal morphological features and color of seeds which will inevitably the main visual factors in seed classification process. Seed identification in certain crops such as wheat, corn, etc. is very significant task because of dissimilarities exists between varietal morphology and quality [3]. Traditional methods for identifying seed varieties that relies on manual inspection is time consuming and requires professional knowledge. In recent years, machine learning and deep learning are the emerging technology for identifying varieties of seeds. The specific objectives of the paper is: (1) to distinguish the wheat varieties; (2) to explore and accomplish classification of different varieties of seeds using different classifier models such as Decision Tree, Random Forest, Nearest Neighbors, Gaussian Process, Neural Net, Ada Boost, Naive Bayes, Support Vector Machine (SVM) Linear, SVM RBF (SVM with the Radial Basis Function) and SVM Sigmoid with 2, 5, 10-fold cross validation; (3) to visualize the classification results evaluated using different classifiers models.

2   Related Work

Seed is a crucial and basic input to obtain high crop yield and sustainable growth in agricultural production. Production of quality seeds and distribution of assured quality seeds is a crucial task in the agricultural field. It is tiresome process which leads to accidental mixing between different variety of wheat seeds, poor quality seeds packaged in the packets and large amount of fine quality seeds wasted during this process. An efficient and automated seed identification and classification, seed growth analysis [6], seed testing and packaging facility is need of an hour to evaluate thousands of seeds. Machine learning technique is an ideal approach for these problems. This would helpful to reject the poor quality seeds and cultivate good quality seeds. Also it ensures packaging of fine quality of seeds [11, 12]. In this section, related works to this study are reviewed and discussed. The main objective of this paper is to propose a model to classify wheat seed varieties. In order to accomplish this task and understand the problem better, our literature review focused on five aspects which is done by different researchers: 1) Seed classification using Machine Learning techniques 2) Classification of rice grains using Neural Networks 3) Autonomous wheat seed type classifier system 4) Agricultural crop yield prediction Using ANN Approach and 5) Agricultural data prediction by means of neural network. [7] states Neural Networks to classify Rice varieties which contain a total of 9 different rice verities. They developed different neural network models to extract thirteen morphological features, six color features and fifteen texture features from color images of individual seed samples. Finally, they combined feature model produced with an overall classification accuracy of 92%. [5] proposed a novel solution to solve the problem faced by the commercial farmers and seed packaging industries and enhance the efficiency in seed cultivation and seed packaging process. [10] suggested a model as a combination of pattern recognition and vision-based approach along with the neural networks to classify the five corn seed varieties. The results exhibited average classification accuracy up and around 90% for five stated varieties of the corn seeds [8].

When comparing the existing system, some shortcomings were found that the performance of related techniques in the literature varies significantly and most researchers have evaluated their method with limited number of datasets. Hence the large variety of seeds must be addressed and robust approaches should be developed to provide feasible solution for the farmers as well as rice seed industry.

3 Proposed Methodology

The Proposed Machine Learning Based Wheat Seed Classification is depicted as a layered Architecture in Fig. 1. The diagram is divided in to four layers. The first layer is an input layer which is used to collect data from UCI website [14]. The next layer is processing layer which is used to split dataset into training and testing dataset. The next classification layer is used to classify three wheat classes using machine learning classifiers. The final layer is the output layer is used to find the accuracy, precision, recall and f1 score for different classifiers for different seeds.

Fig. 1. Proposed system architecture

3.1 Input Layer: Sample Collection and Pre-processing

In order to classify different varieties of wheat seeds, the dataset of wheat seeds is gathered from the great dataset repository UCI website [14]. The number of samples of wheat seeds available in the dataset are 210. Among the different varieties of wheat seeds classes, Kama, Rosa and Canadian are taken for the classification process. Seven geometrical or morphological features of seeds are considered on the basis of which seeds are classified into three classes of wheat.

3.2 Processing Layer

Dataset is split into training and testing dataset

3.3 Classification Layer: Machine Learning Based Classification

The performance was measured using different classifier algorithms with K-fold cross validations. The following describes the different classifier algorithms in Machine Learning.

K-Nearest Neighbours: The K-NN classifies test sample based on the majority of its K-Nearest Neighbors with minimum distance signifies most common attributes. The determination of K is crucial for K-NN. In this study, K was optimized by comparing K-NN models using K from 3 to 100 with a step of 1. Here, K distance was selected as 20 after cross validation.

Naive Bayes Classifier: The Naive Bayes is a statistical classifier which is based on Bayes theorem. This method predicts probabilities of a given samples belonging to a specific class, which means that it provides the probability of occurrence of a given sample or data points within a particular class. The following equation is used to explain the principle of Bayes' theorem:

$$P(H|X) = \frac{P(H|X)P(H)}{P(X)} \tag{1}$$

where $P(H|X)$ is the posterior and $P(H)$ is the prior probability of class (target) whereas $P(X|H)$ and $P(X)$ are the likelihood and prior probabilities of predictor respectively.

Decision Tree: Decision tree has influenced wider area in Machine Learning. It can be used for both classification and prediction. It uses tree like model and used to represent decisions. Its goal is to create a model to predict a value of a target variable. It is a map of possible outcomes from a series of related choices. Decision tree starts with a single node which branches to possible decision. Each of the possible decision leads to additional nodes which branch off into other possibilities.

Random Forest: Random Forest is a popular machine learning algorithm that can be used for both classification and regression problems. It based on the concept of ensemble learning and it belongs to the supervised learning technique contains a number of decision trees on various subsets of the given dataset. It takes the average to improve the predictive accuracy of the dataset. It takes the prediction from each tree and predicts the final output. It leads to higher accuracy and prevents the problem of overfitting when it has greater number of trees.

K-Fold Cross-Validation: K-fold cross validation that makes the data into k equal or equal sized folds or segments or partitions. Partitions of data into k equal or nearly equal sized folds or segments. To being data split into k folds, data is usually stratified earlier. In this process a good representation of data as a whole is confirmed and rearrangement is done if needed. Binary classification problem is an example where each class comprises half of the data. It is great to organize the data such that in every fold, around half of the instances should be in each class.

Ada Boost: Adaptive Boosting is an acronym for Ada Boost algorithm. It is called Adaptive Boosting because the weights are re-assigned to each instance, with higher weights to incorrectly classified instances. Boosting is used to reduce bias as well as the variance for supervised learning. It works on the principle where learners are grown sequentially. Except for the first, each subsequent learner is grown from previously grown learners. In simple words, weak learners are converted into strong ones.

SVM Linear: IT is the most basic type of kernel, usually one dimensional in nature. It proves to be the best function when there are lots of features. The linear kernel is mostly preferred for text-classification problems as most of these kinds of classification problems can be linearly separated. Linear kernel functions are faster than other functions.

$$F(x, xj) = sum(x.xj) \tag{2}$$

Here, **x, xj** represents the data you're trying to classify.

SVM RBF: IT is one of the most preferred and used kernel functions in svm. It is usually chosen for non-linear data. It helps to make proper separation when there is no prior knowledge of data.

$$F(x, xj) = \exp(-gamma * ||x - xj|| \wedge 2) \tag{3}$$

The value of gamma varies from 0 to 1. You have to manually provide the value of gamma in the code. The most preferred value for gamma is 0.1.

SVM Sigmoid: IT is mostly preferred for neural networks. This kernel function is similar to a two-layer perceptron model of the neural network, which works as an activation function for neurons.

$$F(x, xj) = \tanh(\alpha xay + c) \tag{4}$$

Output Layer: Accuracy, precision, recall and F1 score for different classifier algorithms were obtained in this layer.

4 Results and Discussion

In this study, Wheat classes Kama, Rosa and Canadian seeds classified using different classifiers using the dataset that was taken UCI website [14] which is a huge dataset repository. The performance was measured using different classifier algorithms with K-fold cross validations. It is observed that all classifiers using 2-fold cross validation gives highest performance except Neural Net and SVM sigmoid. The Accuracy for the classification of the wheat varieties has been computed using the following formulas which uses numerical details of correctly classified class from total samples of wheat in the dataset (Tables 1 and 2).

$$Accuracy = \frac{no.\ of\ identified\ samples}{total\ no.\ of\ samples} *100 \tag{5}$$

The Precision Recall and F1 Score are also the important measure to consider for system evaluations which are calculated as follows (Table 3):

$$Precision = \frac{\sum True\ Positives}{\sum (True\ Positives + False\ Positives)} * 100 \tag{6}$$

Table 1. Kama seed classification using different classifiers and 2,5,10 fold cross validation

	2 K-fold			5 K-fold			10 K-fold		
	Precision	Recall	F1 Score	Precision	Recall	F1 Score	Precision	Recall	F1 Score
K Nearest Neighbors	1.0	1.0	1.0	1.0	1.0	1.0	1.0	1.0	1.0
Gaussian Process	1.0	1.0	1.0	1.0	1.0	1.0	1.0	1.0	1.0
Decision Tree	1.0	1.0	1.0	1.0	1.0	1.0	1.0	1.0	1.0
Random Forest	1.0	1.0	1.0	1.0	1.0	1.0	1.0	1.0	1.0
Neural Net	1.0	0.9	0.95	1.0	0.9	0.95	1.0	0.2	0.33
Ada Boost	1.0	1.0	1.0	1.0	1.0	1.0	1.0	1.0	1.0
Naive Bayes	1.0	1.0	1.0	1.0	1.0	1.0	1.0	1.0	1.0
SVM Linear	1.0	1.0	1.0	1.0	1.0	1.0	1.0	1.0	1.0
SVM RBF	1.0	1.0	1.0	1.0	1.0	1.0	1.0	1.0	1.0
SVM Sigmoid	1.0	0.5	0.67	1.0	0.5	0.67	1.0	0.5	0.67

Table 2. Rosa seed classification using different classifiers and 2,5,10 fold cross validation

	2 K-fold			5 K-fold			10 K-fold		
	Precision	Recall	F1 Score	Precision	Recall	F1 Score	Precision	Recall	
K Nearest Neighbors	1.0	1.0	1.0	1.0	1.0	1.0	1.0	1.0	
Gaussian Process	1.0	1.0	1.0	1.0	1.0	1.0	1.0	1.0	
Decision Tree	1.0	1.0	1.0	1.0	1.0	1.0	1.0	1.0	
Random Forest	1.0	1.0	1.0	1.0	1.0	1.0	1.0	1.0	
Neural Net	0.88	1.0	0.94	0.88	1.0	0.94	0.38	0.67	
Ada Boost	1.0	1.0	1.0	1.0	1.0	1.0	1.0	1.0	
Naive Bayes	1.0	1.0	1.0	1.0	1.0	1.0	1.0	1.0	
SVM Linear	1.0	1.0	1.0	1.0	1.0	1.0	1.0	1.0	
SVM RBF	1.0	1.0	1.0	1.0	1.0	1.0	1.0	1.0	
SVM Sigmoid	0.1	0.17	0.11	0.1	0.17	0.11	0.1	0.17	

$$\text{Recall} = \frac{\sum \text{True Positives}}{\sum (\text{True Positives} + \text{False Negatives})} * 100 \tag{7}$$

$$\text{F1 Score} = 2 * \frac{\text{Precision} * \text{Recall}}{\text{Precision} + \text{Recall}} \tag{8}$$

Table 3. Canadian seed classification using different classifiers and 2,5,10 fold cross validation

	2 K-fold			5 K-fold			10 K-fold		
	Precision	Recall	F1 Score	Precision	Recall	F1 Score	Precision	Recall	F1 Score
K Nearest Neighbors	1.0	1.0	1.0	1.0	1.0	1.0	1.0	1.0	1.0
Gaussian Process	1.0	1.0	1.0	1.0	1.0	1.0	1.0	1.0	1.0
Decision Tree	1.0	1.0	1.0	1.0	1.0	1.0	1.0	1.0	1.0
Random Forest	1.0	1.0	1.0	1.0	1.0	1.0	1.0	1.0	1.0
Neural Net	1.0	1.0	1.0	1.0	1.0	1.0	0.78	1.0	0.88
Ada Boost	1.0	1.0	1.0	1.0	1.0	1.0	1.0	1.0	1.0
Naive Bayes	1.0	1.0	1.0	1.0	1.0	1.0	1.0	1.0	1.0
SVM Linear	1.0	1.0	1.0	1.0	1.0	1.0	1.0	1.0	1.0
SVM RBF	1.0	1.0	1.0	1.0	1.0	1.0	1.0	1.0	1.0
SVM Sigmoid	0	0	0	0	0	0	0	0	0

The performance was measured using different classifier algorithms with K-fold cross validations. All classifiers using 2-fold cross validation gives highest performance except Neural Net and SVM sigmoid. It is observed that the performance decreases when the number of fold decreases. All the classifiers except Neural Net and SVM sigmoid gives highest performance on 2-fold cross validation. Figure 2 shows the overall Precision, Recall and F1 score for three wheat classes for different classifiers in different k-fold cross validation.

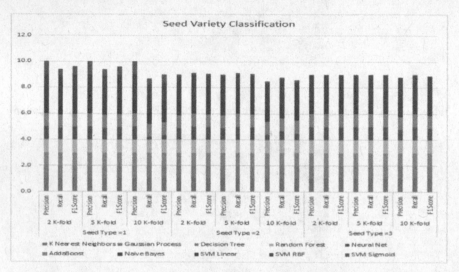

Fig. 2. Seed classification using different classifiers and 2,5,10 fold cross validation

5 Conclusion

In this paper Wheat classes Kama, Rosa and Canadian seeds classified using different classifiers using the dataset that was taken UCI website [14] which is a huge dataset repository. The performance was measured using different classifier algorithms with K-fold cross validations. It is observed that all classifiers using 2-fold cross validation gives highest performance except Neural Net and SVM sigmoid. Also noticed that the performance decreases when the number of fold decreases. All the classifiers except Neural Net and SVM sigmoid gives highest performance on 2-fold cross validation. In future, some other deep learning classifiers used to further improve the performance. Also, there are unsupervised machine learning techniques such as clustering also used to classify the seed varieties. Moreover, other categories of seed varieties also included for classification.

References

1. JayaBrindha, G., Subbu, E.S.G.: Ant colony technique for optimizing the order of cascaded svm classifier for sunflower seed classification. IEEE Tran. Emerg. Top. Comput. Intell. **2**(1), 78–88 (2018). https://doi.org/10.1109/TETCI.2017.2772918

2. Fabiyi, S.D., et al.: Varietal classification of rice seeds using RGB and hyperspectral images. IEEE Access **8**, 22493–22505 (2020). https://doi.org/10.1109/ACCESS.2020.2969847
3. Zapotoczny, P., Zielinska, M., Nita, Z.: Application of image analysis for the varietal classification of barley: morphological features. J. Cereal Sci. **48**(1), 104–110 (2008)
4. Hiremath, S. K., Suresh, S., Kale, S., Ranjana, R., Suma, K. V., Nethra, N.: Seed Segregation using Deep Learning, pp. 1–4. Grace Hopper Celebration India (GHCI), Bangalore, India (2019). https://doi.org/10.1109/GHCI47972.2019.9071810
5. Bhurtel, M., Shrestha, J., Lama, N., Bhattarai, S., Uprety, A., Guragain, M.: Deep Learning based Seed Quality Tester (2019)
6. Scott Howarth, P.C.S.M.: Measurement of seedling growth rate by machine vision, p. 1836–1836–10 (1993). https://doi.org/10.1117/12.144027
7. Silva, C.S., Sonnadara, U.: Classification of Rice Grains Using Neural Networks. In: Proceedings of Technical Sessions, vol. 29, pp. 9–14. Sri Lanka (2013)
8. FAOSTAT: Food and Agricultural Commodities Production. http://www.fao.org/faostat/en/#data/QC (2017). Accessed 1 Jul 2019
9. Šťastný, J., Konečný, V., Trenz, O.: Agricultural data prediction by means of neural network. Agr. Econ. **2011**(7), 356–361 (2011)
10. Chen, X., Xun, Y., Li, W., Zhang, J.: Combining discriminant analysis and neural networks for corn variety identification. Comput. Electron. Agric. **71**, S48–S53 (2010)
11. Ali, A., et al.: Machine learning approach for the classification of corn seed using hybrid features. Int. J. Food Prop. **23**, 1097–1111 (2020)
12. Jamuna, K.S., Karpagavalli, S., Vijaya, M.S., Revathi, P., Gokilavani, S., Madhiya, E.: Classification of Seed Cotton Yield Based on the Growth Stages of Cotton Crop Using Machine Learning Techniques. In: 2010 International Conference on Advances in Computer Engineering, pp. 312–315 (2010). https://doi.org/10.1109/ACE.2010.71
13. FAOSTAT: Food Supply. http://www.fao.org/faostat/en/#data/CC (2013). Accessed 1 Jul 2019
14. https://archive.ics.uci.edu/ml/machine-learning-databases/00236/

Allocation of Overdue Loans in a Sub-Saharan Africa Microfinance Institution

Andreia Araújo[1], Filipe Portela[1]([⊠]), Filipe Alvelos[1], and Saulo Ruiz[2]

[1] Algoritmi Centre, University of Minho, Braga, Portugal
a75270@alunos.uminho.pt, cfp@dsi.uminho.pt, falvelos@dps.uminho.pt
[2] Pelican Rhythms, Porto, Portugal
saulo@pelican-technology.com

Abstract. Microfinance is one strategy followed to provide opportunities to different economic classes of a country. With more loans, there is a high risk of increasing the loans entering the overdue stage, overloading the resources available to take action on the repayment. In this paper, it is only approached the experiment using clustering to the problem. This experiment was focus on a segmentation of the overdue loans in different groups, from where it would be possible to know what loans could be more or less priority. It showed good results, with a clear visualization of three clusters in the data, through Principal Component Analysis (PCA). To reinforce this good visualization, the final silhouette score was 0.194 which reflects that is a model that can be trusted. This way, an implementation of clustering loans into three groups, and a respective prioritization scale would be the best strategy to organize the loans in the team and to assign them in an optimal way to maximize the recovery.

Keywords: Assignment problem · Data mining · Microfinance

1 Introduction

This work is based on an after-stage from the usual loan process. The usual loan process should finish on the thirtieth day after the application was approved (or before if it is paid earlier). Sometimes, the process can take more than usual when the user is not capable of paying on time, entering in an overdue phase. This represents loss and a problem to this MFI.

That overdue phase (before defaulting) can be divided into some different stages and each one has different approaches made to the user during them. The problem approached in this study is in this overdue phase, more concretely in the calls that the officers need to make to the users that did not pay the loan.

Every day, a list of users to call are made and distributed to each officer. The first point on the list is promised calls which means they were called before and

---

This work has been supported by FCT - Fundação para a Ciência e Tecnologia within the R&D Units Project Scope: UIDB/00319/2020, and Pelican Rhythms.

the user provided a date to repay but it was not fulfilled. The second point is loans that changed stages (for example, that finished stage A without payment and entered stage B) or loans that are on their eightieth day of overdue. Next, there are the loans that did not have the opportunity to be called. And, after, the ones that were called but did not have an answer. Lastly, the list covers the loans that are still in the last two stages of overdue. Note that whenever the list achieves 150 loans to be contacted, the order is stopped and the contact list closed. Since this rules are based on a business perspective, the company decided to search deeply on its data to try to optimize the lists.

So, the objective of this project started to be the development of a model of allocation between overdue loans and officers from the MFI Credit Collection Team (CC team), being the main research question the optimization of the allocation of overdue loans to the CC team. And, with that in mind, three experiments were conducted to seek a distribution of the loans through the officers available, to maximize the probability of recovery. The first two were based on the classification technique from Data Mining; and the last one has used the clustering technique. In this paper, it will be only approached the third experiment.

Therefore, it was possible to achieve two important goals from that model with well-defined clusters: the definition of a prioritization scale to call the users from overdue loans, and the definition of the list of contacts to be called by the officers, based on the prioritization scale.

Lastly, to structure the communication of this project into this paper, it was decided to divide into 7 sections: Introduction, Background (where the context of the project is presented), Material and Methods (where some tools are described), Case Study (where it is approached a data characterization and how it was modeled), Discussion of the Results and Conclusion.

2   Background

In this section, a context of the problem is presented, being described why microfinance is important and how an application to a loan proceeds.

2.1   Importance of Microfinance

During the last global recession faced in 2008, one of the systems that struggled more was the banking system. In a country, this has a negative impact on its economy and, on its development.

In Sub-Saharan Africa, the story was not different. Some countries of the region are still facing an economic recovery from their latest recessions, reflecting a lot of challenges to the stability of the banking system. [1] addresses nonperforming loans, capital inadequacy, and non-transparency as three variables of part of the failure in reaching that stability [3].

The focus of this work was only the first point, which [2] refers to as a major cause of the economic failure or even the start of a banking crisis.

To fight poverty and benefit the social balance between economic classes, online microfinance was taken as a solution [4]. Peer-to-peer (P2P) is one of the modalities created and it is responsible for making the connection between private lenders and borrowers in developing countries, through the internet [5,6], offering to entrepreneurs the opportunity to have access to a quicker option of credit.

2.2   Process of an Application to Loan

For the Sub-Saharan microfinance institution (MFI) under study, the investors can invest in the MFI aiming to grow the loan portfolio, but they will not decide which loan their money goes to. This is made automatically when the application is processed and accepted, so the borrowers will receive the amount the moment there is a final decision. The lending process can be defined in these steps:

1. The user must download the app and log in;
2. The user fills in the profile and loan details;
3. The loans are evaluated based on the probability of default by processing all the details gathered from the device;
4. If the loan is accepted, the money is paid out to the bank account from the user directly;
5. The user repays via bank transfer or online transaction using a debit card, within 30 days after the loans get accepted.
6. If the user does not repay on time, there is a grace period of 3 days before being redirected to overdue. There, the Credit Collection Team will contact the user to become aware of the payment and the risks associated.

3   Material and Methods

When starting the research project, it is important to define the most suitable approach and philosophy in order to better formulate the research plan for the research question - how to optimize the allocation of overdue loans to Credit Collection Team officers.

The strategy to follow is Design Science Research (DSR) [14], which focuses on problem-solving, understanding first the problem background and the solutions-oriented through the creation of technological artifacts (such as models) that create an impact on organizations [13].

In order to define the artifact, a data mining process will be done. So, for this reason, a Cross-Industry Standard Process for Data Mining (CRISP-DM) methodology can be followed, which describes an approach to data mining [15].

Regarding the algorithm selection, when considering customer segmentation based on its value, clustering has an advantage on this analysis since it is responsible for organizing the data into similar groups, which makes it easier to build an action plan for each cluster of customers based on the different characteristics each has [11]. [12] also show good results when K-means is applied in a situation of customer segmentation. This is an approach that can be adapted to the context of the problem that is being solved, so K-means will be the algorithm selected.

4   Case Study

In this section, it is described the dataset used in the approach of the analysis presented in the paper and the modeling part.

4.1   Data Structure

The approach presented in this paper is more focused on the loan itself and the behavior of their users, independently of the characteristics of officers. The target was clustering the loans and for these reasons, the dataset used to model this situation (dataset B) has loan-related variables that compose it, with a reinforcement of the number of SMS features.

Since the dataset was quite balanced, there was no need to apply any resampling strategy. Table 1 shows a description of some characteristics that build an overview of dataset B.

Table 1. Description of an overview of dataset B.

	Dataset B
Number of rows	26,825
Period	2019-01-01 to 2019-11-01
Percentage paid	55.50%
Percentage defaulting	44.50%
Total number of variables	56
Features related to SMS	32

4.2   Data Description

The variables collected follow the same basis as in credit scoring models, being related to personal information, loan characteristics [7,8] and mobile phone usage features [9].

Firstly, it is possible to access the personal information of a user from the moment a loan application is settled for the first time. There, the user needs to fill a profile that includes personal information, such as date of birth and professional/ educational background; it includes information related to their family (marital status and number of dependents) and property like owning a car. Some of these variables can be updated later without creating a new profile to continue to apply for loans in this institution.

Secondly, the loan characteristics are added based on the purpose of each application a user makes. In this case, the amount and the reason for the loan are selected as well when filling different applications.

Finally, regarding the mobile phone usage features it was used some features related to the installation of applications (app) - what type of apps the user usually download before loan application and during loan process. And also SMS received during the loan period (which is one of the differences for a credit scoring model) and before the approval of the loan, considering the address and the content of the SMS.

4.3   Modeling

In this modeling phase, the goal expected was to try to find group patterns in the data exported and prepared from dataset B. This way, a clustering method based on the K-Means algorithm was tested to discover if there are groups with similar behavior on the data and if it would be possible to prioritize the loans based on what group they would be part of.

In this case, the target variable considered is a label that indicates which identified group the loan is part of. The definition of these labels is based on the ideal number of clusters, identified here recurring to the Elbow Method [10]. This method provides a k number of clusters, which indicates the target variable varies between 0 and $k-1$. Having in mind that the target variable is the cluster, a loan x could belong to Cluster 0 to $k-1$.

In Fig. 1, it is displayed the Elbow Method with the sum of squared errors and the number of clusters in the axes. It is expected to be found the point where the balance between the sum squared errors and the number of clusters is optimal - the "elbow" of the curve. This point is when the number of clusters is 3, so from now on the aim is to visualize them on the data and characterize them.

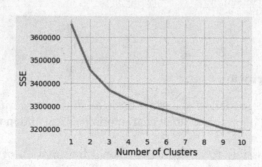

Fig. 1. Optimal number of clusters displayed by the Elbow Method.

Next, after running the K-Means algorithm, it was possible to model the three clusters and their centers. In order to be able to visualize them and evaluate how well they were defined, it was needed to apply the Principal Component Analysis (PCA) method since the number of features modeled was too large to be possible to represent them in two axes graphic.

Then, it is possible to select the center of each cluster and analyze the differences between them and how distinctly they behave. If there are specific characteristics that have a higher impact in a singular cluster when compared to the others, then it is decided to filter the data by these variables and rerun the K-Means algorithm. A new visualization is deployed and it is compared to the previous one.

After these steps, the clusters are analyzed and defined/characterized based on three criteria selected following a business perspective.

Lastly, the result is the definition of the prioritization scale supported by these three criteria.

5   Results

In this section, the results are presented, explained, and analyzed based on the visualization of the clusters using PCA.

5.1   The Experiment

Succeeding the two new equivalent variables to deploy from the application of the PCA, the graph in Fig. 2 was plotted.

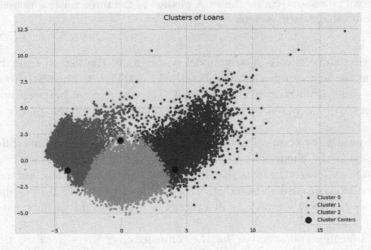

Fig. 2. Visual representation by the PCA method of the three clusters present in this data.

The silhouette score from Fig. 2 was only 0.043, which even if it is higher than 0, is still poor. The reason was related to the fact that the majority of the number of features are common in all clusters, as it could be possible to conclude from an analysis of the centers of each cluster. So, it was decided to resample the initial data containing just the features where it is possible to see big oscillations in the coordinates of the centers of the cluster.

The new subset contains all the variables related to the SMSs and the variable *principal*, which is the amount of money requested by the user.

Therefore, the modeling process was repeated and a new graph of the clusters plotted. The graph is shown in Fig. 3.

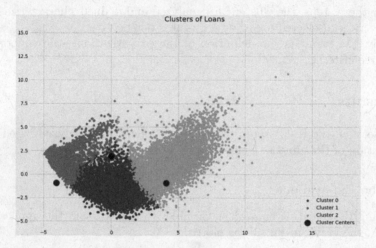

Fig. 3. Visual representation by the PCA method of the three clusters present in the resample after analyse the centers.

Note that since the K-Means model was redone, the legend from Fig. 3 is different as the new number labeled is settled randomly, so cluster 1 from Fig. 2 is not the same from cluster 1 from Fig. 3. The comparison between both representations will be based on the size and the position of the points in each cluster.

Consequently, by comparing both Figs. 2 and 3, there are minimal differences in the shape of the clusters and in the position of the centers, which can verify that these features are the ones that impact the cluster formation. Furthermore, the silhouette score also reflected it, with an improvement to 0.194. Since this score varies between −1 and 1, in practice a 0.194 would mean trust in the model.

From a business perspective, for the company case, there were created three criteria to define which clusters would have the loans that require higher prioritization. These criteria can be described, with the order mattering, as:

1. **Percentage of loans recovered per cluster:** To check where the amount of paid loans is the highest, and consequently with more probability of a return;

2. **Amount of defaulting money per cluster:** To show where the biggest amount of money can be lost is. This results from the sum of the variable *principal*;
3. **Average from a particular SMS feature:** This criterion will not be described in detail to protect the business side of this MFI.

After setting these criteria, it was built a table, Table 2, where the results for each cluster can be consulted. Furthermore, these were analyzed following the importance of each criterion (shown by the order enumerated).

Table 2. Criteria results per each cluster.

Cluster	Percentage of loans recovered	Amount of defaulting	Average of an SMS feature
0	51.77%	124,178,220.00	59,024,711.00
1	50.50%	66,444,430.00	70,482.00
2	67.77%	89,303,870.00	94,007,441.00

For the first criterion, Cluster 2 has the highest percentage whereas Cluster 0 and Cluster 1 are similar. So, there is now some advantage for Cluster 2 in the prioritization scale.

For the second criterion, Cluster 0 represents a higher amount of possible defaulting money, but Cluster 2 shows high values as well. For this reason, Cluster 2 keeps the advantage to be the most prioritized (since the first criterion has more weight on the decision), but Cluster 0 is now settled to be the second in the prioritization scale.

For the third and last criteria, the results came to confirm that Cluster 0 has more potential than Cluster 1.

To sum up, the prioritisation for the loans would go according to the following order: Cluster 2, Cluster 0, and Cluster 1.

5.2 Allocation Model

After labeling the overdue loans, what is still missing is to allocate them to the officers available, following the prioritization rules. In order for that to be possible, an integer programming model was created to optimize the number of loans per officer.

The first action is transforming the output from the clustering model into an input for this model. Thereunto, three variables were created to provide such information:

$$W_i = \begin{cases} 1 & \text{if the loan } i \text{ is part of Cluster 0} \\ 0 & \text{otherwise} \end{cases}$$

$$Y_i = \begin{cases} 1 & \text{if the loan } i \text{ is part of Cluster 1} \\ 0 & \text{otherwise} \end{cases}$$

$$Z_i = \begin{cases} 1 & \text{if the loan } i \text{ is part of Cluster 2} \\ 0 & \text{otherwise} \end{cases}$$

In addition, there is one more piece of information that comes too from the clustering model - the prioritization of the loans. So, three weights were created, that will be associated with each Cluster to add the value matching this prioritization scale, where γ (weight of the loans belonging to Cluster 2) $> \alpha$ (weight of the loans belonging to Cluster 0) $> \beta$ (weight of the loans belonging to Cluster 1).

The goal of this allocation model is to maximize the total weight of the loans assigned to an officer and the mathematical model is:

$$\text{Maximize} \quad \sum_{i=1}^{n} \sum_{j=1}^{m} (\alpha W_i + \beta Y_i + \gamma Z_i) x_{ij}$$

$$\text{Subject to} \quad \sum_{i=1}^{n} x_{ij} = 1, \qquad \text{for each loan } j$$

$$\sum_{j=1}^{m} x_{ij} \le k, \qquad \text{for each officer } i$$

$$x_{ij} \in \{0, 1\}, \qquad \text{for } i = 1, ..., n \text{ and } j = 1, ..., m$$

where $x_{ij} = 1$ if loan j is assigned to officer i, 0 if not. First, the objective function indicates the maximization of the allocation of most priority loans to the officers. Then, the first set of constraints ensures that every loan is assigned to only one officer and the second set of constraints ensures that every officer does not have more assigned loans than their capacity.

6   Conclusion

With the increase of the number of users requesting loans in this MFI, there is a higher risk associated with having loans going overdue, as it would require more human resources available to be capable of handling all the calls needed to influence people to repay.

This MFI decided to take action and revise strategies to analyze, reorganize and improve the performance of the Credit Collection Team. The focus would fall on options to fight the lack of the organization when assigning a loan to officers since there were not many rules and a loan was handled by different officers, always through the different steps, without knowing if it would be profitable or not investing time in that loan.

Summing up all the project, in order to achieve these good results, three scenarios were explored, however only the third one - finding clusters and define with it a prioritization scale - is presented in this paper. In this case, it was used one clustering algorithm - K-Means, where the best results come from, having three clusters well defined by PCA and a final silhouette score of 0.194, which shows a good possibility to rely on it. This way, it was possible to build a prioritization scale: Group 2 > Group 0 > Group 1, based on the rearrangement of the overdue loans into three groups (Cluster 0, Cluster 1, and Cluster 2) from these results, that made this project achieve the purposed goals.

References

1. Adeyemi, B.: Bank failure in Nigeria: a consequence of capital inadequacy, lack of transparency and nonperforming loans. Banks Bank Syst. **6**(1), 99–109 (2011)
2. Reinhart, C.M., Rogoff, K.S.: From financial crash to debt crisis. Am. Econ. Rev. **101**(5), 1676–1706 (2011)
3. Mpofu, T.R., Nikolaidou, E.: Determinants of credit risk in the banking system in Sub-Saharan Africa. Rev. Dev. Financ. **8**(2), 141–153 (2018)
4. Yang, L., Wang, Z., Ding, Y., Hahn, J.: The role of online peer-to-peer lending in crisis response: evidence from Kiva (2016)
5. McIntosh, C.: Monitoring repayment in online peer-to-peer lending (2011)
6. Flannery, M.: Kiva and the birth of person-to-person microfinance. Innov. Technol. Gov. Globalization **2**(1–2), 31–56 (2007)
7. Huang, C.L., Chen, M.C., Wang, C.J.: Credit scoring with a data mining approach based on support vector machines. Expert Syst. Appl. **33**(4), 847–856 (2007)
8. West, D.: Neural network credit scoring models. Comput. Oper. Res. **27**(11–12), 1131–1152 (2000)
9. Björkegren, D., Grissen, D.: Behavior revealed in mobile phone usage predicts loan repayment (2018)
10. Grus, J.: Data Science from Scratch: First Principles with Python. O'Reilly Media (2019)
11. Ngai, E.W., Xiu, L., Chau, D.C.: Application of data mining techniques in customer relationship management: a literature review and classification. Expert Syst. Appl. **36**(2), 2592–2602 (2009)
12. Cheng, C.H., Chen, Y.S.: Classifying the segmentation of customer value via RFM model and RS theory. Expert Syst. Appl. **36**(3), 4176–4184 (2009)
13. Hevner, A.R., March, S.T., Park, J., Ram, S.: Design science in information systems research. MIS Q. 75–105 (2004)
14. Peffers, K., Tuunanen, T., Rothenberger, M.A., Chatterjee, S.: A design science research methodology for information systems research. J. Manag. Inf. Syst. **24**(3), 45–77 (2007)
15. Olson, D.L., Delen, D.: Advanced Data Mining Techniques. Springer, Heidelberg (2008)

Automatic Model for Relation Extraction from Text Documents Using Deep Learning Neural Network

B. Lavanya$^{(\boxtimes)}$ (iD) and G. Sasipriya (iD)

Department of Computer Science, University of Madras, Chennai, India
lavanmu@gmail.com

Abstract. Relation extraction is a significant stage in the information extraction process as it establishes the semantic relation between entities. In nanoparticles, relation extraction is vital, especially in nanomedicine, where the benefits are substantial. The necessity for automatic relation extraction in the field of nanoparticles is an imminent need, as human data extraction is nearly impossible due to the ever-growing number of scientific papers on this subject. Advanced technologies and new research approaches to improve and refine the properties of nanoparticles have emerged in recent years. Our aim is to train and build a model for the automatic extraction of relations from research articles. This paper summarizes how a framework was developed for a relation extraction system for nanoparticles by collecting information from existing research articles, training the model, and using the Natural Language Processing model on the relations of extracted text. An open-source Natural Language Processing model framework formed the basis for our implementation, and it received a significant F1 score for each threshold.

Keywords: Nanoparticles · Automatic extraction · Natural Language Processing · Relation extraction

1 Introduction

In recent years, the field of Nanoinformatics has seen significant growth in observational data. Regardless of publications, the essential information is available in unstructured text data. It is rather difficult to transform these unstructured data into structured data manually. Hence the need for automatic text relation extraction in converting them to the machine-understandable structured dataset format is imminent. The slightest changes in the properties of nanoparticles pose a threat to both humans and the environment. Nanoinformatics is primarily implemented in the realm of medicine to amend and aspire for top-notch prediction models [17]. The task to read and infer relations between the entities from the research articles is simplified with the help of Natural Language

Supported by University of Madras and DST-RUSA 2.0.

Processing. Automatic text relation extraction for unstructured data has been explored in bioinformatics over the last decade, notably for protein and gene, drug diseases [1] between bacteria and location entities [4], and so on.

In recent years, various relation extraction datasets have been created in the field of bioinformatics [18], but it is budding in the field of nanoparticles and is receiving extensive research attention. The major issue for the relation extraction is that contains several named entity recognitions are asymmetrical because of their relation types, possible relation instances should be considered in a sentence for every entity pair. To overcome the shortcoming of unstable predictability of the relation between the entity pairs, there is a need for a neural network model.

This study represents the proposed work in generating the training model for relation extraction in nanoparticles properties and their applications. The input text is very unstructured, so extended the proposed model to generate automatic input data for relation extraction model training. This paper reviews the relation extraction in the field of nanoparticles with good precision and recall. The next part of this paper talks about the relation extraction and neural network for the relation extraction, architecture of proposed approach, training model, results, and discussions.

2   Related Works

This section represents the basic methods of extraction methods for entity relationships and about the basic text mining methods like NEL, NER and REL.

2.1   Extraction Methods for Entity Relationships: Classification

The intention of automatic information extraction is to achieve useful information from the innumerous amount of literature, which if done manually is almost impossible. Basically, the text mining system comprises named entity recognition (NER), named entity linking (NEL), and relation extraction (RE) tasks [15]. NER is modeled that can be undertaken by assigning a tag to each word in a sentence. NER will talk about the words and about the type of the word.

Named Entity Recognition: The named entity recognition is a subtask to identify and classify named entities in an unstructured text. The sequence of the task in NER begins with the sentence by splitting into tokens and then labeling them into categories like part of speech (POS tagging) and entity extraction.

Named Entity Linking: NEL is undertaken to link the entities mentions in the text with their related entities in the knowledge base. NEL has a wide range of applications other than the information extraction and used in Information Retrieval, Content Analysis, Intelligent Tagging, Question Answering System, Recommender Systems, etc. the entity linking system is broadly classified into three modules, namely Candidate Entity Generation, Candidate Entity Ranking, and Unlinkable Mention Prediction [14].

Relation Extraction: Automatic relation retrieval capability is highly useful in fields such nanomaterials arena, where there is an imminent need to detect in the mammoth data critical information, which is hiding in the enormity of the data itself. Common types of ML algorithms are supervised learning, unsupervised learning, and semi-supervised learning [20]. The computing method of sentence processing in supervised learning has been broadly classified into the kernel-based method and Eigenvector-based method, the continuous process of the supervised method is deep learning. Support vector machines (SVM), bootstrap, and remote monitoring approaches are examples of semi-supervised learning methods [6,11].

3   Deep Neural Network-Relation Extraction

Deep Learning methods are becoming essential owing to their clearly show the existence to the accomplishment of an aim at difficult task for complex learning problems [8]. Nevertheless, increasing computing resources at high performance and open-source libraries are making life easier for researchers in the field of bioinformatics and nano informatics [5].

Neural networks have been extensively used in natural language processing, various languages, and image processing and also in some other fields and getting very worthy results [16]. Many of the researchers used deep learning techniques for biomedical data, and the traditional deep learning methods are SVM, Max-Ent, RNN, MV-RNN, CNN+softmax, etc. In the past few years, deep learning techniques [15], for relation extraction is one of the key natural language processing tasks which used RNN and also gives excellent results. Deep learning's success in biomedical NLP is attributed in part to the creation of word vector models such Word2vector [9], ELMO [10], BERT [3], GPT [12], transformer-xl [2], GPT-2 [13].

Deep Learning also helps in word embeddings from word vectors to syntactic and semantic word connections. Standard RNNs can be replaced by Long Short - Term Memory. LSTM networks are ideal for biological literature as they deal with more complex sentences. After the simple introduction to the neural network for the deep learning model, let us have brief look at basic model concepts, that are so essential and come across these models.

3.1   Word Embeddings

Deep learning can help in building an in-depth relation extraction model to strengthen the extraction performance. One of the key prerequisites of deep learning is establishing the semantic relationship information in word to vector format. Word representation helps achieve the same. Word embedding is sufficiently good notice, that it is able to train the quality of vectors by using a simple and also an understanding a neural network model.

3.2   Position Embedding

The absolute positions from 1 to the maximum sequence length are encoded via position embedding, in most cases the value is 512. Each position, in another respect, has a learnable embedding vector. The absolute position embedding is used to model how a token in one position connects with a token in another position.

In the work of relation extraction, the relation between the entities is possessing to the word close to the target entities that are usually interesting information. The position embedding can be specified by entity pairs. The CNN helps to track to find each word that is close to parent or child entities, and also it defines the merging from the current word to parent or child entities using the relative distance [7].

3.3   Extraction Models for Entities and Relationships

In deep neural network models, it is possible to avoid the complex manual features by replacing effective entities and relations features, and it also provides a better effect. The main concept of extracting entities and their relation are classified into types: pipeline models and joint models. Both the Named entity recognition (NER) and relation extraction are giving attention to this pipeline models. By simultaneously joint model will extract entities and relations [19].

Pipeline Models and Joint Models. Machine learning workflows are automated using a machine learning pipeline. They work by allowing a series of data to be converted and associated in a model that can be tested and reviewed to produce a positive or negative result. CNN will extract the keyword information and sustain the information in the sentence. LSTM, the model uses different variants and capturing long sequences of long-range dependencies, successfully bring out the aim in text mining and analysis. To study the features of given sentences and the undertaken quality to make full use of the proposed joint model-based deep neural network.

4   Proposed Model

This section represents the basic structure of system architecture, the joint task of entities and relations such as tagger, parser, relation extractor and token to vector.

4.1   System Architecture

The system architecture of the proposed relation extraction system has the following key steps. The first step is to extract the unstructured text from research articles present in pdf format, cleaning the sentences and extracting them. This is followed by extracting the sentences with named entities in the articles. The very next step is to parse and process them using the model developed for relation extraction, which is described in the next section. The following Fig. 1 represents the architecture of relation extraction.

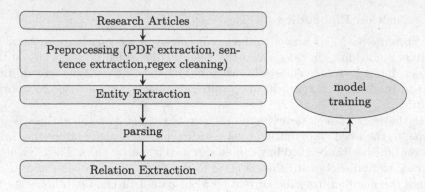

Fig. 1. Overall architecture of relation extraction

4.2   Model Training

For the system to identify relations between two entities, First, the entities need to be extracted, then need to find appropriate POS and dependency tags for all the tokens in the sentence. As they involve many steps, the system needs to have multiple stages to do these works. As in any deep learning NLP framework, our model also incorporates all these steps. Each and every stage is called a pipeline. The different pipelines used to train the model are tagger, parser, lemmatizer, attribute ruler, ner, tok2vec and relation extractor [4].

The tagger part is used to tag for the part of speech (POS). They help in understanding the context of the word. The parser is the pipeline, which is responsible for the dependency parsing of the sentence. The root word is estimated and the relations with other children are depicted using this. The lemmatizer does the job of lemmatizing the words. Often overlooked lemmatized words may result in inefficient tagging and parsing.

An attribute ruler is a pipe that is responsible for rule base pattern matching in the sentence. NER is the most important pipeline of any NLP framework. The entities are tagged using appropriate labels in this pipeline. The following Fig. 2 represents the overall architecture of the training model. For relation extraction, the token to vector and deep learning-based relation extraction is used to train the data by using python NLP framework. For a token, to vector hash embedding-based CNN algorithm is used [7]. For relation extraction deep learning with Adam optimizer is used.

Training is the most important step of any NLP model. The efficient the training is, the model is better with good accuracy. For training the model, the entities and their positions are needed in the document and the parent-child relation between the entities. As the text in the nano informatics domain is mostly unstructured, the first challenge we faced is to automate the token start and end positions in the sentence. For achieving this, the BEIOU methodology has been used. Here B stands for Begin of a multi-token entity, I for an Intermediate token of a multi-token entity, and E for End token of a multi-token entity. O stands for Outside of an entity and U stands for a Single token entity.

The input the user needs to give is a json file that contains information about the text, span tokens, parent and child relationship, entity label, and relation label. The json file is parsed and iterated each and every token to check if is part of an entity span mentioned in the json file. An NLP document object has been created, which is used to find the start and end index of the matching tokens from entity spans. In deep learning based neural network based models, the popular combinations of training, validation, and test data is 80, 10, 10. In this implementation, the 80, 10, 10 ratio produced favorable performance. The common training parameters are optimized and, used in this implementation (batch size - 2, number of epochs - 250, dropout - 0.1), which resulted in an efficient model. The dataset was subjected to 5-fold cross-validation, and the findings appear to be promising. Cross-fold validations are better measured than single-fold validation.

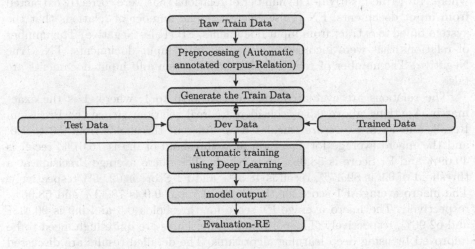

Fig. 2. Proposed model architecture of relation extraction

5   Results and Discussion

5.1   System Implementation

The unstructured text is extracted from PDF files. Then they are processed using our NER model designed for extracting properties and values of nanoparticles, which uses machine learning and rule-based entity extraction. The output containing the relevant sentences with properties and values with the application. These sentences are now processed with our relation extraction model. The entries with relation more than 0.5 are normally considered real relations with an approximate match, above 0.9 are close to strict match and above 0.99 are considered as a strict match.

The total number of sentences extracted from the research articles is 1866. After cleaning headers, footers, email addresses, and references, the number of sentences extracted that are relevant for nanoparticle property and its value with some relations is 54. The number of entities extracted from the extracted sentences is 107. The number of strictly matched relations is 72, while the number of approximately matched relations is 333.

5.2   Performance Measures

As always in any NLP system, Precision, Recall and F1-Score metrics are used to measure the performance of the model [15]. The formula to calculate Precision, Recall and F1 measure are discussed below.

$$Precision = \frac{TP}{TP + FP} Recall = \frac{TP}{TP + FN} F1 - Score = \frac{2PR}{precision + recall} \tag{1}$$

where, TP (True Positive)-The number of relations that were correctly extracted from input documents. FN (False Negative)-The number of relations that the system failed to extract from input documents. FP (False Negative)-The number of relations that were incorrectly extracted from input documents. TN (True Negative)-The number of relations that both system and input documents are false.

The relations are evaluated with a score of 0 to 1, where, 1 is the exact match. Anything above 0.5 is considered a real relation. Hence the Precision, Recall, and F1 measure are calculated on different thresholds from 0.0 to 1.0, and the macro average for Precision at a threshold of 0.9 is 76.67%, recall is 49.09% and F1 Score is 58.06% respectively. The micro average Precision at a threshold of 0.9 is 89.47%, recall is 48.57% and F1 Score is 62.96% respectively. The macro average F1 score for thresholds 0.5 and 0.9 is 78.43% and 58.06%, respectively. The micro average F1 score for thresholds 0.5 and 0.9 is 90.41% and 62.96%, respectively. The performance measures are quite high, need to be improved by using deep learning algorithms. The detailed results are discussed below in Table 1 and Table 2.

Table 1. Training model-macro average

Threshold	Precision	Recall	F1 Score
0.0	5.24	100.00	9.69
0.05	47.75	83.33	55.26
0.1	55.40	83.33	63.22
0.2	64.06	83.33	70.20
0.3	70.48	83.33	74.81
0.4	71.98	83.33	75.91
0.5	75.76	83.33	78.43
0.6	76.67	83.33	79.17
0.7	75.93	78.17	76.50
0.8	76.19	68.25	70.26
0.9	76.67	49.09	58.06

Table 2. Training model-micro average

Threshold	Precision	Recall	F1score
0.0	3.21	100.00	6.21
0.05	31.73	94.29	47.48
0.1	46.48	94.29	62.26
0.2	60.00	94.29	73.33
0.3	73.33	94.29	82.50
0.4	76.74	94.29	84.62
0.5	86.84	94.29	90.41
0.6	89.19	94.29	91.67
0.7	88.57	88.57	88.57
0.8	90.00	77.14	83.08
0.9	89.47	48.57	62.96

6   Conclusion

In this paper, an automatic annotation model is explored to extract the relations between the nanoparticle property and their values. The proposed training model employs automated input for the deep learning model, which can benefit the processing of a greater number of research papers. Using a deep learning-based model to extract entity relations for nanoparticle research articles saves time and cost. According to our findings, deep neural networks may create competitive advantages results with less annotation of input data and less support from external resources such as knowledge bases. Our approach will enhance nanoparticle text mining research, specifically in terms of extracting particle property values and their relationships. The recall value is considerably lesser than the precision value, which needs to be improved. Integration of entity recognition and relation extraction into a unified model with high precision and recall is another task that remains to be investigated.

Acknowledgement. This research work was supported by DST-RUSA through University of Madras under RUSA 2.0 project.

References

1. Ben Abdessalem Karaa, W., Alkhammash, E.H., Bchir, A.: Drug disease relation extraction from biomedical literature using NLP and machine learning. Mob. Inf. Syst. **2021** (2021)
2. Dai, Z., Yang, Z., Yang, Y., Carbonell, J., Le, Q.V., Salakhutdinov, R.: Transformer-XL: attentive language models beyond a fixed-length context. arXiv preprint arXiv:1901.02860 (2019)

3. Devlin, J., Chang, M.W., Lee, K., Toutanova, K.: BERT: pre-training of deep bidirectional transformers for language understanding. arXiv preprint arXiv:1810.04805 (2018)

4. Li, F., Zhang, M., Fu, G., Ji, D.: A neural joint model for entity and relation extraction from biomedical text. BMC Bioinform. **18**(1), 1–11 (2017)

5. Li, J., Sun, A., Han, J., Li, C.: A survey on deep learning for named entity recognition. IEEE Trans. Knowl. Data Eng. (2020)

6. Li, Z., Qu, L., Xu, Q., Johnson, M.: Unsupervised pre-training with Seq2Seq reconstruction loss for deep relation extraction models. In: Proceedings of the Australasian Language Technology Association Workshop 2016, pp. 54–64 (2016)

7. Lin, Y., Shen, S., Liu, Z., Luan, H., Sun, M.: Neural relation extraction with selective attention over instances. In: Proceedings of the 54th Annual Meeting of the Association for Computational Linguistics (Volume 1: Long Papers), pp. 2124–2133 (2016)

8. Lopez, M.M., Kalita, J.: Deep learning applied to NLP. arXiv preprint arXiv:1703.03091 (2017)

9. Mikolov, T., Sutskever, I., Chen, K., Corrado, G.S., Dean, J.: Distributed representations of words and phrases and their compositionality. In: Advances in Neural Information Processing Systems, pp. 3111–3119 (2013)

10. Peters, M.E., et al.: Deep contextualized word representations. arXiv preprint arXiv:1802.05365 (2018)

11. Quan, C., Wang, M., Ren, F.: An unsupervised text mining method for relation extraction from biomedical literature. PLoS ONE **9**(7), e102039 (2014)

12. Radford, A., Narasimhan, K., Salimans, T., Sutskever, I.: Improving language understanding by generative pre-training (2018)

13. Radford, A., Wu, J., Child, R., Luan, D., Amodei, D., Sutskever, I., et al.: Language models are unsupervised multitask learners. OpenAI blog **1**(8), 9 (2019)

14. Shen, W., Wang, J., Han, J.: Entity linking with a knowledge base: issues, techniques, and solutions. IEEE Trans. Knowl. Data Eng. **27**(2), 443–460 (2014)

15. Sousa, D., Lamurias, A., Couto, F.M.: Using neural networks for relation extraction from biomedical literature. In: Cartwright, H. (ed.) Artificial Neural Networks. MMB, vol. 2190, pp. 289–305. Springer, New York (2021). https://doi.org/10.1007/978-1-0716-0826-5_14

16. Wang, H., Qin, K., Zakari, R.Y., Lu, G., Yin, J.: Deep neural network based relation extraction: an overview. arXiv preprint arXiv:2101.01907 (2021)

17. Xiao, L., Tang, K., Liu, X., Yang, H., Chen, Z., Xu, R.: Information extraction from nanotoxicity related publications. In: 2013 IEEE International Conference on Bioinformatics and Biomedicine, pp. 25–30. IEEE (2013)

18. Xing, R., Luo, J., Song, T.: BioRel: towards large-scale biomedical relation extraction. BMC Bioinform. **21**(16), 1–13 (2020)

19. Xue, L., Qing, S., Pengzhou, Z.: Relation extraction based on deep learning. In: 2018 IEEE/ACIS 17th International Conference on Computer and Information Science (ICIS), pp. 687–691. IEEE (2018)

20. Zhang, Q., Chen, M., Liu, L.: A review on entity relation extraction. In: 2017 Second International Conference on Mechanical, Control and Computer Engineering (ICMCCE), pp. 178–183. IEEE (2017)

Polarized Extractive Summarization of Online Product Reviews

Gendeti Manjju Shree Devy(✉) ⓘ, Korupolu Saideepthi(✉) ⓘ,
Varakala Sowmya(✉) ⓘ, and Rajendra Prasath(✉) ⓘ

Indian Institute of Information Technology Sri City, 630 Gnan Marg, Chittoor,
Sri City 517 646, Andhra Pradesh, India
{manjjushreedevy.g18,saideepthi.k18,sowmya.v18,rajendra.prasath}@iiits.in,
drrprasath@gmail.com
http://rajendra.2power3.com

Abstract. In the e-commerce domain, online customer reviews offer valuable information to manufacturers and potential buyers. Many fast-moving products receive hundreds or even thousands of online reviews. Therefore, summarizing multiple online reviews makes it easier for merchants and buyers to establish the usefulness of products and to prioritize the issues that are most important to them. Motivated by this, we designed an automated two-step pipeline for Polarized Extractive Summarization that uses Machine Learning and Deep Learning models. The first module of the pipeline is trained on a manually annotated Amazon cellphone dataset that contains multiple reviews of verified purchases on Amazon. Polarized Sentiments from these reviews are extracted by splitting the review(s) into sentences. The second module removes the duplicates and near-duplicates; the selected sentences are given to the graph-based model so as to rank these sentences based on the Link Analysis. The top-ranked sentences are chosen to generate positively and negatively polarized summaries. We evaluate our proposed models on manually annotated Amazon review dataset for sentence classification as well as summarization which yields better results on both modules. This research sets out to identify the sentiments in each review that contributes towards the summarization and removal of near duplicates. It has been shown from the experimental results that the generated summaries are more accurate and informative.

Keywords: Classification · Online reviews · Data mining · Deep neural networks · Near duplicate detection · Extractive summarization · Polarized summarization

1 Introduction

The development of e-commerce has led to an explosion of thousands of reviews, opinions, and comments all over the world in recent years. When consumers review products online, they discuss all aspects of it. As a result of reviews, manufacturers can learn what kind of critics/comments/suggestions related to the products that

ⓒ Springer Nature Switzerland AG 2022
R. Chbeir et al. (Eds.): MIKE 2021, LNAI 13119, pp. 147–160, 2022.
https://doi.org/10.1007/978-3-031-21517-9_15

customers have already purchased. It also benefits a new buyer to get to know the features, pros and cons of the products. However, it can also be very challenging for the customers to go through hundreds or thousands of reviews available online for a particular product. Online review data is so large and manually analysing the content is impractical, time-consuming and effort-intensive. So it is necessary to apply automatic summarization of the review content.

In the modern world, text summarization is an important task of extracting essential information from a large sized text documents and creating a simplified and informative version of the entire document for the reader [7,14,21]. Summarization of text has become increasingly popular in the past decade [2,19,22,27]. Generally, text summarization can be divided into single-text and multi-text summarizations [10]. A single-document summarization described in [18] addresses the problem of graph-based keyword extraction for the summarization of a single document. A single text summarization thus does not conflict with the opinions voiced within it. In addition, the effect of novelty of content is not considered in single-text summarization because a single document is released at a specific time point. Comparatively, multiple-text summarization processes multiple documents within a single subject [8,11]. When dealing with multi-text summarizations, it is important to resolve the conflict between certain opinions raised by different authors in addition to managing semantic expression consistency in the summarization results.

Research works reported in [3,31] primarily summarize online reviews with feature based, but not accounting the opinion of the users. The summaries are gradually given in accordance with the user's opinion, where the work proposed by Wang et al. [29] shows that the summaries on the basis of the customer's perception and experience about the features. Abdi et al. [1] proposed an approach that is a hybrid deep learning architecture for opinion-oriented multi-document summarization based on multi-feature fusion. The work proposed by Hou et al. [13] considered five different aspects about the features of a product namely emotion, perceptions, product affordance, usage conditions and product features to generate summaries.

The purpose of this paper is to provide two polarized summaries, since a single summary based on multiple reviews would have conflicts of user's opinions. It is important to find out the sentiment behind the reviews in order to develop polarized summaries, but product reviews can also relate to both positive and negative aspects. So we split up the reviews into sentences and then determine their sentiment based on their polarity.

The paper is organized as follows: Sect. 2 presents works related to different types of summarization. The dataset is described in Sect. 4 and Sect. 5 discusses the proposed methodology. The experimental results are discussed in Sect. 6. Lastly, Sect. 7 provides conclusions, limitations, and pointers to future work.

2   Related Work

In this section, we briefly present past research works, describing the pros and cons of each approach.

El-Kassa *et al.* [5] describes different types of summarization techniques and a detailed survey of summarization approaches. They have given a fundamental layout of strategies that can be utilized for summarization. Text summarization can be categorized into different types. Two popular approaches are *extractive summarization* [30] and *abstractive summarization* [15]. Extractive summarization creates the summary from phrases or sentences from the source input. The abstractive summarization approach represents the input text in an intermediate form and then generates the summary with words and sentences that differ from the original text sentences. Most studies focus on different techniques of extractive summarization [26,30] since the abstractive summarization needs an extensive NLP tasks.

Text summarization is classified based on the nature of the yield summary as Generic [9] and Query-Based [28]. A generic summarizer extracts more imperative data by preserving the generic subject of the given content and also it represents the view of the authors. A query-based summary presents the information that is most important to the initial search query. It is also to be noted that a generic summary gives a general sense of the content. The query-based summary is very often alluded to as a query-focused, topic-focused, or user-focused summary [7].

There are numerous works on extractive text summarization for online reviews [26,30]. A feature-based summarization [12] where for each product, a summary has been created utilizing Phrasal Extraction-Parts-of-Speech (POS) tagging and topic modeling - *Latent Dirichlet Allocation* (LDA). A graph-based approach proposed in [20] uses cosine similarity and modified TextRank. All the above-mentioned works focus on generating a single summary, but we aim to provide two summaries based on user reviews in which pros about a product will be included in a positive summary and flaws/cons will be considered vital for a negative summary. In order to achieve this task, we need to perform sentiment classification of online reviews at the sentence level in prior.

Over the decades, there has been an immense improvement in sentence classification and there are different classification methods on online reviews. Yoon Kim [16] proposed a Convolutional Neural Network (CNN) based approach for sentence classification in which experiments on different combinations of word embeddings and different CNN layers were illustrated. There are a few works that concentrate on sentiment based summarization, opinion-oriented multi-document summarization [1] is an approach in which sentiment classification using RNN-LSTM algorithm and sentence ranking using RBM algorithm occurs simultaneously to classify a sentence and on top of it sentence selection is performed to generate summaries. Sentiment Lossless Summarization [17] is a graph-based extractive summarizer which configures the input text as a weighted directed graph with scores as edges of the graph then sentiment of each node and sentiment compensation weights are calculated to maintain the sentiment consistency between the candidate summaries and the original text.

The summarization approaches described in [1,17] focus on the entire review and also some works [6] portray the sentiment of review based on the ratings as well. But the 5-star rated reviews may contain the negative aspects and a 1-star rated reviews may contain a positive point of a product in some cases. Hence, we opted to do a sentence-wise sentiment detection instead of the entire review to make the summary more exact and precise in terms of its polarity.

3   Motivation and Contributions

All the above mentioned works in Sect. 2 focus on the comments of the users and different aspects of the products for generating a single summary, where all these summaries highlight either only the features of a product or a generic summary of a product alone. But for buyers to understand the pros and cons and the manufactures to understand user requirements of the product, we need incorporate the positive and negative aspects of various features in summaries. In order to come up with two summaries related to opposite polarities, we ought to perform sentiment classification with aspect-based approaches. The previous works including the one described in [6] concentrate on the entire review alone and describe the sentiment of review based on the rating but in most of the cases the reviews contain multiple sentiment sentences with positive, negative and neutral polarities. So, it is essential to propose an approach that identifies sentiment of the sentences in reviews and also generates informative polarized summaries.

The contributions of our work can be summarized as follows:

(1) To the best of our knowledge, this work is the first of this kind that generates two summaries which are positive-opinionated and negative-opinionated. This work performs sentence-level information extraction to predict sentiment polarity and selects an informative sentence to generate generic extractive summaries on Amazon reviews.
(2) We have applied different techniques to handle challenges: (i) multiple sentiments in the same review; (ii) Rewriting "TextSpeak" words; (iii) Rewriting contractions; (iv) Converting emojis, and then the preprocessed data is given to sentence classification module.
(3) Summarization: After a substantial amount of research on summarization, to make summaries more optimize, we additionally performed near-duplicates detection and removal, where the unique sentences will be retained to the graph-based summarization module to generate precise summaries.
(4) At the end, we conducted extensive experiments to test the viability of our proposed method. We have reported our results on the dataset that we are proposing: the Amazon Cellphone Reviews Dataset[1] which is on amazon cell phone reviews. The proposed method outperforms the existing methods on the standard metrics, and the statistically significant results make it a suitable method for the polarized summarization of online product reviews.

---

[1] Dataset: https://github.com/manju1201/Polarized-Summarization-of-Reviews.

4   Dataset

We have created a new dataset because the datasets available online are on complete reviews, and a review on a whole can not be given a single polarity because the review may contain both positive and negative aspects of user experiences. With the goal of facilitating a more precise polarised summaries, we are preparing a dataset containing sentence classifications as positive, negative, and neutral. A sample user review and the sentences extracted from it with different polarities are given in Table 1.

Table 1. Sample user review and the polarities associated with it

Sample User Review	Sentences with Polarity	
	Polarity	Extracted Polarized Sentences
I've been using this for the last 1 month. This is a good phone as far as speed is concerned. I bought this phone about a month back and now the primary camera has stopped focussing.	Positive	This is a good phone as far as speed is concerned.
	Neutral	I've been using this for the last 1 month.
	Negative	I bought this phone about a month back and now the primary camera has stopped focussing

4.1   Amazon Cellphone Reviews Dataset

This dataset was obtained by crawling the phone reviews of verified users on Amazon's website, extracting the review portion of the content and splitting them into 12,000 sentences. Six popular brands such as Apple, Samsung, Oppo, Vivo, Redmi, and OnePlus were considered. In each brand, we have captured 5 different configurations and crawled 100 reviews. In order to create a coherent dataset, the reviews were preprocessed by separating them by the "full stop" (.) and manually annotated by the polarity of the sentences. Our dataset contains the following distribution of polarized sentences: *Positive Sentences*(5982), *Negative Sentences*(4138), and *Neutral Sentences*(1880). We are collecting and adding new pairs of review-sentences with sentiments to this dataset periodically.

5   Methodology

We propose a new approach for generating summaries that use sentiment analysis to address the shortcomings of the existing approaches discussed in Sect. 2. In this work, GloVe word embedding, deep learning models like the BERT model, machine learning models like k−means and graph based models like Page Rank are used to better capture data and produce informative and efficient summaries.

5.1 Preprocessing

Data preprocessing is applied to convert the raw data into a machine-readable format. We often encounter incomplete, inconsistent, redundant, and noisy data in online reviews.

Algorithm 1. Proposed TSWR algorithm

Input: Basic Preprocessed Sentence
Output: TSWR Preprocessed Sentence
$words = [], alternative = [], processed\_words = []$
$words[] = sentence.Split("")$
for k to $len(words)$ **do**
   $alternative[k] \leftarrow$ wordsreplacement(words [k], textspeakwords)
   if alternative[k] not null **then**
     $processed\_words[k] \leftarrow$ alternative[k]
   else
     $processed\_words[k] \leftarrow$ words[k]
   end if
end for
for k to $len(processed\_words)$ **do**
   $TSWR\_preprocessed\_text$ $+=$ $processed\_words[k]+$ " "
end for
return $TSWR\_preprocessed\_text$

There are several steps involved in preprocessing raw data so that it can be transformed into a comprehensible format. Preprocessing steps include (i) replacing the emojis to text Ex: "💜" is converted into text format as "purple heart" and (ii) decontracted the contractions like "don't" to "do not" and (iii) normalised the data. While annotating, we came across a few words which are short forms or text language words, we call them "textspeak words" Eg: "gud" for "good", "grt" for "great", which we collected manually and stored in a dictionary. We implemented a new algorithm Text Speak Words Replacement (TSWR), which replaced the "textspeak" words with most suitable words. We did not remove the stop words neither we did lemmatization because if we do these steps the final points in summary are not readable.

5.2 Sentiment Classification - Module A

Instead of considering the whole review, we created a dataset where a review is split into sentences, and a sentiment is measured for each sentence using sentiment classification models. We implemented basic machine learning models and some deep learning models. We found that the dataset had imbalanced data, so we resampled it to ensure that the labels were in equal proportion and we also performed class weighting (Fig. 1).

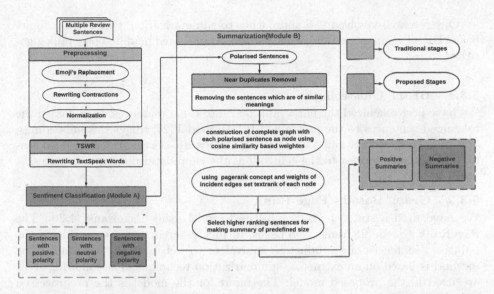

Fig. 1. The pipeline showing polarized summarization by removal of near duplicates

In our approach, the first step is to create word embeddings that transform sentences into input matrices. Pre-trained GloVe embeddings[2] were used to generate the input matrix of the preprocessed data, with the input being the maximum sentence length. The next most important step is constructing the neural network layers where three parallel layers of 2 dimensional convolution with filter lengths of 3, 4 and 5 as well as max pool layers for each convolutional layer are applied to the output of the embedding layer to process sequences of arbitrary length. We next extracted the sentiment of a sentence from the obtained vectors, and for better results we also applied various regularization techniques, which will be discussed in experiments.

As a result of the sentiment classification, we get a dataset with predicted sentences with positive, negative and neutral polarities. These data are now inputs for summarization models.

5.3   Summarization - Module B

Now let us consider the review sentences of a particular product. All predicted positive sentences are concatenated with a full stop and saved as a string. Assume that the concatenated string is a bin. Each product has a positive, negative and neutral bin. The bins are fed into a summarization model, which generates the top 15 sentences of the product. As a result, a product has top 15 positive sentences which would be useful for buyers who are willing to purchase it, and top 15 negative sentences would be helpful for manufacturers as suggestions.

---

[2] Pre-trained GloVe embeddings: https://nlp.stanford.edu/projects/glove/.

Our research involved two approaches to summarization. One being a topic-based BERT summarizer and other being a graph-based modified pagerank algorithm.

5.3.1   BERT Clustering

We have first tokenized the paragraph text into clean sentences, then passed the tokenized sentences to the pretrained BERT model [24] to get the embeddings, and then clustered the embeddings using $k-$means, selecting the embedded sentences that were closest to the centroid as the candidate summary sentences.

5.3.2   Graph Based - Page Rank

We have used a state of art graph-based model using PageRank [4, 20]. The PageRank system [23] ranks web pages. In this research, we have considered the polarized sentences of a product as equivalent to web pages in PageRank. The method is based on an extractive summarization technique. The following steps are used in the proposed model: The input for the model is the preprocessed polarized sentences.

Step 1: Construction of cosine similarity based weighted graph: PageRank links(similar) important sentences with other important sentences. In order to define similarity between input sentences, we represent each sentence as a word vector. Google Word2Vec was used to obtain vector representations of these input sentences. Vector length in Google word2vec is about 300 features and these features are learnt using unsupervised machine learning models. Next, a complete weighted graph $G = (V, E, W)$ is built using Networkx Library[3] where V is the set of vertices, each of which corresponds to a sentence represented by the word vector; the set E represents the edges in the graph and the set W represents weights assigned with the graph edges. Using the sentence vectors which act as vertices in the graph, a weighted edges are generated using cosine similarity of the 2 sentence vectors of the graph.

Step 2: Ranking sentences: We have a graph $G = (V, E, W)$ with V as the set of vertices, E as the set of edges, and W as the set of weights. Each node in this graph is assigned a TextRank based on the weights of incident edges and PageRank. We now have ranks for each sentence vector.

Step 3: Selecting n sentences: After obtaining the PageRank rankings for the sentences, the vertices of the graph are arranged in descending order by their PageRank scores. Then, for summarization, the top n scored sentences are selected (n is a user-defined value, 15 is considered). The summary is constructed from the top n sentences. However, the model has a drawback. This summarization model uses sentences from multiple reviews with a single polarity, i.e. sentences that have the same sense but different styles. In the summary,

---

[3] Networkx Library: https://github.com/networkx/networkx,.

these sentences end up appearing if they have high rank. In order to remove duplicate entries, we changed the model as described below.

Step 4: Remove Near Duplicates: Near duplicates are those sentences which have similar meaning but are written in a different way. With the graph-based model, every sentence in the document is compared with every other sentence and the sentence with the most similarities is the most important sentence. This leads to an important problem where two sentences with the same meaning could have the same importance, and both sentences will end up being in the summary. Therefore, it is important to remove the sentences that are near duplicates.

We implemented two methods to remove near duplicates like **Hashing** and **Cosine-similarity with a threshold**. The first step in hashing is to split each sentence into trigrams and then each trigram is assigned a hash value. Sentences with the same hash values are considered as near-duplicates. The sentences having less information are not given for summarization. In cosine similarity, every sentence is compared with all other sentences and the sentences having more than 60% similarity score are treated as near duplicate sentences.

6   Experiments and Results

As described in Sect. 3, our proposed method is based on sentiment classification and a generic text summarization approach. We therefore employ both sentiment analysis and text summarization in our methodology. This study aims to train a graph-based text summarization model to generate two polarized summaries using the sentence-sentiment pairs dataset. In this section, we discuss the experiments and results of our experiments on sentiment classification and text summarization.

Experiment 1: Sentiment Classification
We performed Random Forest, Logistic regression, Support Vector Machines (SVM) with both TF-IDF and Count-Vectorizer but we could not get the results with an improvement, so we turned to deep learning models. Our dataset was trained on Convolutional Neural Networks with GloVe embeddings under various regularization such as early stopping on validation loss and different dropout layers in order to get better accuracy.

Experiment 2: Summarization
Initially we experimented with BERT pre-trained word embeddings and $k-$means clustering to summarise data, but found those results not adequate for our dataset, so we moved on to graph based models. In Graph-based Text summarization with Text Rank, we additionally included the removal of near duplicates.

Experiment 3: Hashing

First, we used hashing to remove near duplicates. In this method, we first generate tri-grams of the sentences and each tri-gram is given a hash value. For example, S1 : "phone is great i love it" is converted as {S1 : {phone is great : Ha}, {is great i : Hb }, {great i love : Hc}, {i love it : Hd} and S2 : "phone is great" is converted as {S2 : {phone is great : Ha}. As S2 hash values are within S1 these are near duplicates and since S2 is a redundant, it was discarded. However, the drawback to this method is that the dataset we use was collected from real-world users, thus each person has a different style of writing and even though we understand that the meaning of the sentences is the same, but the machine can not understand these sentences. For example S1: this is the best phone i ever used in my life S2: this is a great phone and one of the best i have ever used. These sentences while performing hashing do not get same hash values.

Experiment 4: Cosine similarity with a threshold

Due to the drawback in hashing method, we have used a cosine similarity method to identify near duplicates. After experimenting with different threshold values with cosine similarity for removing near duplicates, we found that 60% threshold provided better results for our dataset. For Example, S1: this is the best phone i ever used in my life, S2: this is a great phone and one of the best i have ever used. The similarity score between these sentences is 64.5% and are considered as near duplicate sentences.

6.1   Performance Metrics

To quantitatively evaluate the performance of our proposed models, we have used the standard measures for classification and summarization.

(i) Classification Metrics

Precision(P), Recall(R) and F_1 measures are calculated, which are widely used for evaluating in sentence classification tasks.

$$\text{Precision}(P) = \frac{\text{True Positives}}{\text{True Positives} + \text{False Positives}} \tag{1}$$

$$\text{Recall}(R) = \frac{\text{True Positives}}{\text{True Positives} + \text{False Negatives}} \tag{2}$$

(ii) Summarization Metrics

We use a set of metrics from ROUGE (Recall-Oriented Understudy for Gisting Evaluation) like Rouge-N [25] and Average Rouge Score (ARS) [1]. Rouge-N is an n-gram recall between a reference summary and candidate summary.

$$\text{ROUGE} - N = \frac{\sum_{S \in \text{reference\_summaries}} \sum_{N-\text{grams}} \text{Count}_{\text{match}}(N - \text{gram})}{\sum_{S \in \text{reference\_summaries}} \sum_{N-\text{grams}} \text{Count}(N - \text{gram})} \tag{3}$$

$$\text{Recall (R)} = \frac{|S_{\text{ref}} \cap S_{cand}|}{|S_{\text{ref}}|} \tag{4}$$

$$\text{Precision(P)} = \frac{|S_{\text{ref}} \cap S_{cand}|}{|S_{\text{cand}}|} \tag{5}$$

F_1 score for both classification and summarization is calculated as follows:

$$F_1 = \frac{2 \cdot \text{Precision} \cdot \text{Recall}}{\text{Precision} + \text{Recall}} \tag{6}$$

6.2 Results and Discussion

While machine learning models have improved, deep learning models like convolutional neural networks have started yielding better results (Table 2).

Table 2. Results on sentence-classification, all reported metrics are weighted averages.

Model A	Recall	Precision	F_1-score	Accuracy
Random Forest - Countvectorizer	0.80	0.81	0.81	0.80
SGDClassifier Logistic Regression - Countvectorizer	0.81	0.81	0.81	0.80
SGDClassifier SVM - Countvectorizer	0.80	0.80	0.80	0.79
Random Forest - TF-IDF	0.80	0.80	0.79	0.80
SGDClassifier Logistic Regression TF-IDF	0.81	0.81	0.81	0.81
SGDClassifier SVM TF-IDF	0.81	0.81	0.81	0.80
CNN - GloVe embeddings	**0.86**	**0.86**	**0.86**	**0.86**

Graph-Based-Text-Summarization gives better results than the BERT model because, we can get a better view of important sentences by constructing the similarity graph of sentences in comparison to the $k-$means centroid approach. Our proposed model removed near duplicates, and thus the summary has more unique points than the model that only considers similarities directly from the input sentences (Table 3).

Table 3. Results on summarization models.

Model	BERT-based			Graph-based			Proposed Model		
Sentiment	Positive			Positive			Positive		
Metric	Recall	Precision	F_1-score	Recall	Precision	F_1-score	Recall	Precision	F_1-score
Rouge-1	0.33	0.64	0.44	0.68	0.70	0.69	**0.77**	**0.72**	**0.74**
Rouge-2	0.18	0.34	0.23	0.57	0.57	0.57	**0.65**	**0.62**	**0.63**
ARS	0.36	0.49	0.33	0.62	0.63	0.63	**0.71**	**0.67**	**0.68**
Sentiment	Negative			Negative			Negative		
Metric	Recall	Precision	F_1-score	Recall	Precision	F_1-score	Recall	Precision	F_1-score
Rouge-1	0.41	0.63	0.50	0.74	0.64	0.68	**0.77**	**0.65**	**0.70**
Rouge-2	0.24	0.37	0.29	0.59	0.51	0.55	**0.62**	**0.53**	**0.57**
ARS	0.32	0.50	0.39	0.66	0.57	0.61	**0.69**	**0.59**	**0.63**

7   Conclusion and Future Work

In this paper, we discussed different types of text summarizations of online reviews. Our proposed method does separation of positive and negative points from a review to produce polarized summaries by constructing similarity graphs of sentences after removing duplicates and near duplicates which gives a better view of unique and important sentences than the general graph-based approaches. Furthermore in the prepossessing, the pronouns replacement and the count of emojis can also be done as a factor for the model to decide the sentiment of a sentence and the cosine-similarity to remove near duplicates can be replaced with any advanced methods for finding similarity to improve the model performance. Aspect based summarization can also be attempted for. In conclusion, our approach gives both positive and negative summaries of products, with the positive summaries assisting the customer in buying the product and the negative summaries helping the customer and manufacturer in understanding the flaws.

References

1. Abdi, A., Hasan, S., Shamsuddin, S.M., Idris, N., Piran, J.: A hybrid deep learning architecture for opinion-oriented multi-document summarization based on multi-feature fusion. Knowl.-Based Syst. **213**, 106658 (2021)
2. Abdi, A., Idris, N., Alguliev, R.M., Aliguliyev, R.M.: Automatic summarization assessment through a combination of semantic and syntactic information for intelligent educational systems. Inf. Process. Manage. **51**(4), 340–358 (2015)
3. Bafna, K., Toshniwal, D.: Feature based summarization of customers' reviews of online products. Procedia Comput. Sci. **22**, 142–151 (2013)
4. Brin, S., Page, L.: The anatomy of a large-scale hypertextual web search engine. In: Proceedings of the Seventh International Conference on World Wide Web, no. 7, pp. 107–117. WWW7, Elsevier Science Publishers B.V., NLD (1998)
5. El-Kassas, W.S., Salama, C.R., Rafea, A.A., Mohamed, H.K.: Automatic text summarization: a comprehensive survey. Expert Syst. Appl. **165**, 113679 (2021)

6. Fang, X., Zhan, J.: Sentiment analysis using product review data. J. Big Data **2**(1), 1–14 (2015)

7. Gambhir, M., Gupta, V.: Recent automatic text summarization techniques: a survey. Artif. Intell. Rev. **47**(1), 1–66 (2017)

8. Goldstein, J., Mittal, V.O., Carbonell, J.G., Kantrowitz, M.: Multi-document summarization by sentence extraction. In: NAACL-ANLP 2000 Workshop: Automatic Summarization (2000)

9. Gong, Y., Liu, X.: Generic text summarization using relevance measure and latent semantic analysis. In: Proceedings of the 24th Annual International ACM SIGIR Conference on Research and Development in Information Retrieval, pp. 19–25 (2001)

10. Gupta, V., Lehal, G.S.: A survey of text summarization extractive techniques. J. Emer. Technol. Web Intell. **2**(3), 258–268 (2010)

11. Heu, J.U., Qasim, I., Lee, D.H.: Fodosu: multi-document summarization exploiting semantic analysis based on social folksonomy. Inf. Process. Manage. **51**(1), 212–225 (2015)

12. Hong, M., Wang, H.: Research on customer opinion summarization using topic mining and deep neural network. Math. Comput. Simul. **185**, 88–114 (2021)

13. Hou, T., Yannou, B., Leroy, Y., Poirson, E.: Mining customer product reviews for product development: a summarization process. Exp. Syst. Appl. **132**, 141–150 (2019)

14. Kar, M., Nunes, S., Ribeiro, C.: Summarization of changes in dynamic text collections using latent dirichlet allocation model. Inf. Process. Manag. **51**(6), 809–833 (2015)

15. Khan, A., Salim, N.: A review on abstractive summarization methods. J. Theoret. Appl. Inf. Technol. **59**(1), 64–72 (2014)

16. Kim, Y.: Convolutional neural networks for sentence classification. In: Proceedings of the 2014 Conference on Empirical Methods in Natural Language Processing (EMNLP), pp. 1746–1751. Association for Computational Linguistics, Doha (2014). https://doi.org/10.3115/v1/D14-1181

17. Li, X., Wu, P., Zou, C., Xie, H., Wang, F.L.: Sentiment lossless summarization. Knowl.-Based Syst. **227**, 107170 (2021)

18. Litvak, M., Last, M.: Graph-based keyword extraction for single-document summarization. In: Coling 2008: Proceedings of the Workshop Multi-source Multilingual Information Extraction and Summarization, pp. 17–24 (2008)

19. Ly, D.K., Sugiyama, K., Lin, Z., Kan, M.Y.: Product review summarization from a deeper perspective. In: Proceedings of the 11th Annual International ACM/IEEE Joint Conference on Digital Libraries, pp. 311–314 (2011)

20. Mallick, C., Das, A.K., Dutta, M., Das, A.K., Sarkar, A.: Graph-based text summarization using modified TextRank. In: Nayak, J., Abraham, A., Krishna, B.M., Chandra Sekhar, G.T., Das, A.K. (eds.) Soft Computing in Data Analytics. AISC, vol. 758, pp. 137–146. Springer, Singapore (2019). https://doi.org/10.1007/978-981-13-0514-6_14

21. Mani, I., Maybury, M.T.: Automatic summarization. In: ACL, 39th Annual Meeting and 10th Conference of the European Chapter, Companion Volume to the Proceedings of the Conference: Proceedings of the Student Research Workshop and Tutorial Abstracts, p. 5, 9–11 July 2001, Toulouse, France. CNRS, Toulose, France (2001)

22. Mehta, P.: Survey on movie rating and review summarization in mobile environment. Int. J. Eng. Res. Technol. **2**(3) (2013)

23. Mihalcea, R.: Graph-based ranking algorithms for sentence extraction, applied to text summarization. In: Proceedings of the ACL Interactive Poster and Demonstration Sessions, pp. 170–173 (2004)

24. Miller, D.: Leveraging BERT for extractive text summarization on lectures. arXiv preprint arXiv:1906.04165 (2019)

25. Moratanch, N., Chitrakala, S.: A survey on extractive text summarization. In: 2017 International Conference on Computer, Communication and Signal Processing (ICCCSP), pp. 1–6. IEEE (2017)

26. Nazari, N., Mahdavi, M.: A survey on automatic text summarization. J. AI Data Mining **7**(1), 121–135 (2019)

27. Sankarasubramaniam, Y., Ramanathan, K., Ghosh, S.: Text summarization using Wikipedia. Inf. Process. Manage. **50**(3), 443–461 (2014)

28. Tang, J., Yao, L., Chen, D.: Multi-topic based query-oriented summarization. In: Proceedings of the 2009 SIAM International Conference on Data Mining, pp. 1148–1159. SIAM (2009)

29. Wang, W.M., Li, Z., Tian, Z., Wang, J., Cheng, M.: Extracting and summarizing affective features and responses from online product descriptions and reviews: a kansei text mining approach. Eng. Appl. Artif. Intell. **73**, 149–162 (2018)

30. Wong, K.F., Wu, M., Li, W.: Extractive summarization using supervised and semi-supervised learning. In: Proceedings of the 22nd International Conference on Computational Linguistics (Coling 2008), pp. 985–992 (2008)

31. Xu, X., Meng, T., Cheng, X.: Aspect-based extractive summarization of online reviews. In: Proceedings of the 2011 ACM Symposium on Applied Computing, pp. 968–975 (2011)

Deep Learning Based NLP Embedding Approach for Biosequence Classification

Shamika Ganesan, S. Sachin Kumar$^{(\boxtimes)}$, and K.P. Soman

Centre for Computational Engineering and Networking (CEN),
Amrita School of Engineering, Amrita Vishwa Vidyapeetham,
Coimbatore, Tamil Nadu, India
s_sachinkumar@cb.amrita.edu, kp_soman@amrita.edu

Abstract. Biological sequence analysis involves the study of structural characteristics and chemical composition of a sequence. From a computational perspective, the goal is to represent sequences using vectors which bring out the essential features of the virus and enable efficient classification. Methods such as one-hot encoding, Word2Vec models, etc. have been explored for embedding sequences into the Euclidean plane. But these methods either fail to capture similarity information between k-mers or face the challenge of handling Out-of-Vocabulary (OOV) k-mers. In order to overcome these challenges, in this paper we aim explore the possibility of embedding Biosequences of MERS, SARS and SARS-CoV-2 using Global Vectors (GloVe) model and FastText n-gram representation. We conduct an extensive study to evaluate their performance using classical Machine Learning algorithms and Deep Learning methods. We compare our results with dna2vec, which is an existing Word2Vec approach. Experimental results show that FastText n-gram based sequence embeddings enable deeper insights into understanding the composition of each virus and thus give a classification accuracy close to 1. We also provide a study regarding the patterns in the viruses and support our results using various visualization techniques.

Keywords: Biosequence classification · MERS · SARS · SARS-CoV-2 · NLP · FastText · Global vectors

1 Introduction

Severe Acute Respiratory Syndrome Coronavirus 2 (SARS-CoV-2), the novel Coronavirus belongs to the β-Cov genus of the Coronaviridae family. They are a part of single-stranded RNA virus families. As reported by the World Health Organization (WHO), since the inception of this virus in December 2019, several strains of the virus have been reported in humans worldwide [1]. Analysis of SARS-CoV-2 sequences attains paramount importance in tracking its mutations as well as for vaccine production.

Viruses are inert by nature. They assume power only when attached to the host cells. RNA sequences consist of four bases, Adenine (A), Guanine (G),

© Springer Nature Switzerland AG 2022
R. Chbeir et al. (Eds.): MIKE 2021, LNAI 13119, pp. 161–173, 2022.
https://doi.org/10.1007/978-3-031-21517-9_16

Cytosine (C), Uracil (U). In DNA sequences, Uracil is replaced by Thymine (T). These bases combine with each other using Hydrogen bonds and these bonded bases form a base pair [3]. In RNA sequences like the Coronavirus, these base pairs do not exist since they are single stranded by structure.

The Coronaviridae family consists of four genera viz., α-Cov, β-Cov, γ-Cov and δ-Cov. Among these genera, SARS-CoV-2 belongs to the β-Cov genus. The discovery of SARS virus dates back to November 2002 whereas the Middle East Respiratory Syndrome (MERS) was first reported in 2012. The MERS, SARS and SARS-Cov-2 viruses belong to the Coronaviridae family's β-Cov genus, but they differ in the subgenus. While SARS-CoV-2 belongs to Sarbecovirus subgenus, MERS and SARS belong to Merbecovirus subgenus. Since they are closely related, yet differ in their effects on the human body and the course of medical treatment [4], an efficient method for sequence representation is essential.

After the commencement of the Covid-19 pandemic, several variants of the SARS-CoV-2 virus, have been found worldwide. Understanding the characteristics of this virus and finding common patterns in its chemical composition holds paramount importance. Over time, the viral spread has also observed double mutation and triple mutation variants [5] which require a generalised envelope for detection. Since SARS-CoV-2 sequences share similarities with other correlated sequences, sequence representations play a vital role. In this paper, we explore different computational methods for representing MERS, SARS and SARS-CoV-2 sequences using Natural Language Processing (NLP) based embeddings viz., FastText and Global Vectors (GloVe) and experiment with different parameters to identify the most effective representations by visualizing the results using t-distributed stochastic neighbor embeddings(t-SNE) plots, frequency distribution plots and word cloud representations. We classify the sequences obtained from the different embedding schemes using standard Machine Learning and Deep Learning algorithms to verify the efficiency of the embeddings. The subsequent sections of the paper are organized as follows. A study on existing sequence representation methods are described in Sect. 2. In Sect. 3, various sequence embedding schemes are discussed in more detail. The experimental results are presented in Sect. 4, and Sect. 5 summarizes the paper along with a discussion on future scope.

2   Related Works

The fundamental Bioinformatics study for any virus includes sequence alignment, structural comparison, and molecular docking [12]. Biosequence structural comparison has been carried out using various methods in the existing research including Machine Learning and Deep Learning techniques. Representing sequences as images using the models provided by FeigLab, Zhang group and AlphaFold had been one of the early methods explored in [13]. The use of sequence data in the form of strings has mainly been carried out using one-hot encoding method where either each of the bases are encoded or its k-mers [14,15]. This technique has helped in representing distinct features. However, this method is not capable of understanding correlations between bases or k-mers. This method was extended into a Deep Learning based approach where features were extracted using Convolutional Neural Networks and those extracted features were utilized to identify viruses [16].

In [17], word2vec based models were used for deriving the correct feauture space for SARS-CoV-2 virus sequences. Several embedding sizes were experimented with, to obtain the optimal hyperparameter values. Specialized word2vec models for biosequences were also trained and extensively used in the literature [8,18,19]. For classifying the promoter region of a DNA, [20] took an approach of continuous FastText n-grams. The embeddings obtained were fed to a Convolutional Neural Network (CNN) model for classification. In our work, we attempt to utilize both FastText and GloVe techniques for sequence embeddings.

3   Methodology

The fundamental course of action required for sequence analysis is to map the sequences into a numerical plane. This mapping allows all the sequences and tokens to be embedded into numerical vectors of uniform dimension and capture hidden information like similarities and semantic relationships between k-mers. In this paper, we majorly use two methods of embeddings viz., FastText and Global Vectors (GloVe).

3.1   FastText

FastText[1] is an open source library by Facebook Research, written in C++, that enables unsupervised embeddings. The central idea behind this approach was to modify the Skip-gram method by providing a unique word vector to each word and ignore their internal structure [6].

$$\sum_{p=1}^{P}[\sum_{c \subset C_p} log(1 + e^{-s(w_p w_c)}) + \sum_{n \subset N_{p,c}} log(1 + e^{s(w_p n)})] \tag{1}$$

3.2   Global Vectors (GloVe)

Global Vector (GloVe)[2] is an unsupervised algorithm for representing words in the form of vectors. GloVe models are focused on the concept that word representations should capture similarity information between words along with capturing the statistical information.

4   Proposed Architecture

4.1   Dataset Description

The experiments in this paper have been conducted using the virus sequences obtained from the National Centre for Biotechnology Information (NCBI) Virus

---

[1] https://fasttext.cc/.
[2] https://nlp.stanford.edu/projects/glove/.

databank [7]. The following taxid's were used - SARS-CoV-2 (2697049), SARS (694009) and MERS (1335626). The files were obtained in FASTA format which is a text based format for obtaining sequences of bases, where each of the base is represented using its corresponding single letter. Total number of sequences used for the experiments were 864, 804 and 1691 respectively for SARS-CoV-2, MERS and SARS. Table 1 provides a brief description of the dataset.

4.2   Proposed Experimental Design

Sequence Embedding. The dataset obtained after pre-processing are to be embedded as vectors into the Euclidean plane. This is a pivotal step in sequence analysis, since the obtained embeddings would determine the amount of information captured from the original sequences. Since biological sequences are subject to mutations i.e. variations in the structure of genes which could include introduction of new patterns, deletion of existing patterns, rearrangement of structures and so on, it is extremely crucial for the embedding method to be able to incorporate the following - 1. contextual knowledge of neighbouring patterns 2. ability to incorporate new tokens apart from the training vocabulary. Keeping these factors as the primary focus of our research, in this paper we experiment with different word embedding techniques commonly used in NLP viz. GloVe and FastText, which have been explained in detail in the previous section.

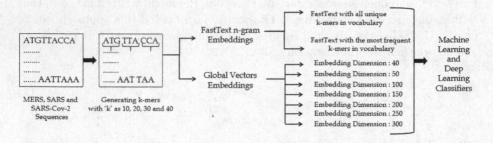

Fig. 1. Block diagram for proposed architecture

The first step is to create k-mers from the sequences. k-mers are tokens generated from any sequence using a shifting window of length 'k'. We have experimented with 'k' values equal to 10, 20, 30 and 40 with a non-overlapping window. For GloVe model, various embedding dimensions were included in the range of 50 to 300. For FastText model, all the embeddings were generated for a fixed dimension of 100. While training the FastText model, two different methods were experimented with - 1. training the model with all unique k-mers in the vocabulary 2. training the model with the most frequent k-mers in the vocabulary. Each of these models were trained for 500 epochs. The FastText embeddings were trained with a minimum n-gram count as 1, since biosequences exist in various lengths, thereby leaving a possibility of unigrams during k-mer generation. The vectors generated for each k-mer are avergaed to generate the overall sequence embeddings.

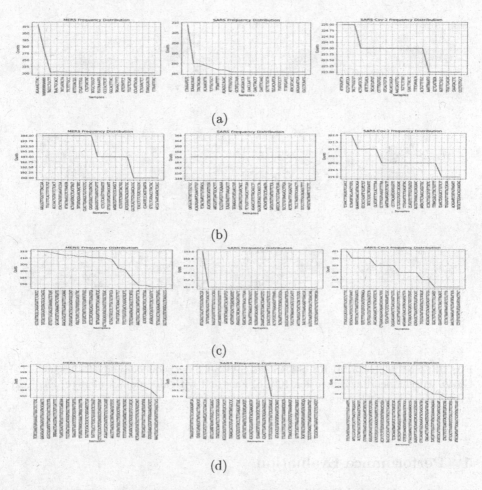

(a)

(b)

(c)

(d)

Fig. 2. Frequency Distribution plots for MERS, SARS and SARS-CoV-2 with (a) 10-mers (b) 20-mers (c) 30-mers (d) 40-mers

Sequence Classification. From a biosequencing perspective, the evaluation of sequence embeddings is carried out primarily using sequence classification. In this paper, we have classified the sequences using Naive Bayes, Decision Trees, Random Forests, Suport Vector Machines (Linear, Polynomial, RBF), AdaBoost, 1D-CNN and 1D-CNN with Bidirectional LSTMs. For each of the Machine Learning algorithms, a 5-fold cross validation was carried out. For the Deep Learning based classifiers, a train-test split of 80-20 was considered using the scikit-learn Machine Learning Library[3]. The results are discussed in Table 2 and 3 and compared with dna2vec [8] in Table 4.

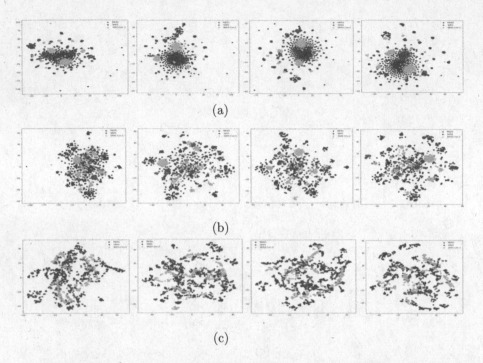

(a)

(b)

(c)

Fig. 3. t-SNE plots for MERS, SARS and SARS-CoV-2 sequence embeddings using (a) GloVe with embedding dimension 100 (b) FastText with all unique k-mers (c) FastText with most frequent k-mers

5   Performance Evaluation

5.1   Experimental Results

For this study, the experiments and analyses were conducted using the Google Colaboratory platform[4] with a hosted runtime of 12GB RAM support. We obtain sequence embeddings using FastText, GloVe and dna2vec models and classify them using various Machine Learning and Deep Learning algorithms as shown in Fig. 1. The cross validation scores and accuracy values for each of them are provided in Table 1, 2 and 3. In order to derive patterns and insights into the sequences and their k-mers, we conduct an in-depth study of these sequences using Frequency Distribution plots, t-SNE plots, Word Cloud representations and ROC curves for visualizing the accuracy for deep learning models.

---

[3] https://scikit-learn.org/stable/.

[4] https://colab.research.google.com/notebooks/intro.ipynb.

Table 1. Cross-validation scores for Machine Learning classifiers and Accuracy for CNN [9] and CNN-BiLSTM [10] with FastText n-gram embeddings

(a) Naive Bayes

k	10	20	30	40
FastText	0.664	0.760	0.907	0.950
FastText with most frequent k-mers	1.000	1.000	1.000	1.000

(b) Decision Tree

k	10	20	30	40
FastText	1.000	1.000	1.000	1.000
FastText with most frequent k-mers	1.000	1.000	1.000	1.000

(c) Support Vector Machine Classifier (Linear Kernel)

k	10	20	30	40
FastText	0.961	0.947	0.990	0.998
FastText with most frequent k-mers	1.000	1.000	1.000	1.000

(d) Support Vector Machine Classifier (Polynomial Kernel)

k	10	20	30	40
FastText	0.265	0.277	0.545	0.433
FastText with most frequent k-mers	1.000	1.000	1.000	1.000

(e) Support Vector Machine Classifier (RBF Kernel)

k	10	20	30	40
FastText	0.934	0.932	1.000	0.933
FastText with most frequent k-mers	1.000	1.000	1.000	1.000

(f) Linear Support Vector Machine Classifier

k	10	20	30	40
FastText	1.000	1.000	1.000	1.000
FastText with most frequent k-mers	1.000	1.000	1.000	1.000

(g) Random Forest

k	10	20	30	40
FastText	1.000	1.000	1.000	1.000
FastText with most frequent k-mers	1.000	1.000	1.000	1.000

(h) AdaBoost

k	10	20	30	40
FastText	0.977	0.990	0.967	0.983
FastText with most frequent k-mers	1.000	1.000	1.000	1.000

(i) CNN [9]

k	10	20	30	40
FastText	0.995	0.997	0.998	0.998
FastText with most frequent k-mers	1.000	1.000	1.000	1.000

(j) CNN-BiLSTM [10]

k	10	20	30	40
FastText	0.985	0.991	0.997	0.992
FastText with most frequent k-mers	1.000	1.000	1.000	1.000

Table 2. Cross-validation scores for Machine Learning classifiers and Accuracy for CNN [9] and CNN-BiLSTM [10] with GloVe embeddings

(a) Naive Bayes

k	10	20	30	40
40	0.552	0.705	0.718	0.685
50	0.548	0.557	0.706	0.702
100	0.557	0.716	0.728	0.686
150	0.569	0.721	0.712	0.706
200	0.534	0.718	0.697	0.698
250	0.538	0.726	0.714	0.693
300	0.548	0.726	0.737	0.707

(b) Decision Tree

k	10	20	30	40
40	0.921	0.910	0.868	0.864
50	0.962	0.938	0.917	0.847
100	0.875	0.917	0.934	0.875
150	0.961	0.880	0.912	0.874
200	0.936	0.904	0.898	0.911
250	0.967	0.871	0.918	0.887
300	0.943	0.912	0.888	0.875

(c) Support Vector Machine Classifier (Linear Kernel)

k	10	20	30	40
40	0.350	0.461	0.429	0.358
50	0.334	0.429	0.426	0.455
100	0.370	0.420	0.363	0.384
150	0.318	0.442	0.364	0.291
200	0.345	0.474	0.415	0.388
250	0.360	0.499	0.403	0.315
300	0.349	0.420	0.438	0.420

(d) Support Vector Machine Classifier (Polynomial Kernel)

k	10	20	30	40
40	0.316	0.270	0.266	0.287
50	0.309	0.263	0.274	0.284
100	0.315	0.341	0.323	0.338
150	0.273	0.263	0.260	0.274
200	0.263	0.358	0.415	0.3405
250	0.300	0.260	0.280	0.295
300	0.358	0.261	0.267	0.293

(continued)

Table 2. (*continued*)

(e) Support Vector Machine Classifier (RBF Kernel)			
k	10	20	30	40
40	0.261	0.210	0.241	0.264
50	0.274	0.247	0.257	0.245
100	0.282	0.249	0.238	0.263
150	0.268	0.268	0.259	0.246
200	0.273	0.257	0.246	0.340
250	0.262	0.261	0.256	0.253
300	0.266	0.262	0.256	0.256

(f) Linear Support Vector Machine Classifier			
k	10	20	30	40
40	0.735	0.772	0.769	0.786
50	0.764	0.776	0.774	0.784
100	0.786	0.785	0.785	0.785
150	0.792	0.783	0.782	0.785
200	0.791	0.783	0.787	0.784
250	0.791	0.783	0.784	0.783
300	0.792	0.783	0.785	0.785

(g) Random Forest			
k	10	20	30	40
40	1.000	1.000	1.000	1.000
50	1.000	1.000	1.000	1.000
100	1.000	1.000	1.000	1.000
150	1.000	1.000	1.000	1.000
200	1.000	1.000	1.000	1.000
250	1.000	1.000	1.000	1.000
300	1.000	1.000	1.000	1.000

(h) AdaBoost			
k	10	20	30	40
40	0.695	0.745	0.803	0.824
50	0.823	0.817	0.819	0.810
100	0.878	0.811	0.809	0.822
150	0.841	0.828	0.815	0.874
200	0.727	0.877	0.819	0.826
250	0.807	0.887	0.856	0.857
300	0.897	0.846	0.695	0.823

(*continued*)

Table 2. (*continued*)

(i) CNN [9]				
k	10	20	30	40
40	0.919	0.894	0.880	0.867
50	0.909	0.882	0.876	0.857
100	0.909	0.885	0.907	0.882
150	0.934	0.889	0.877	0.880
200	0.916	0.915	0.858	0.870
250	0.909	0.897	0.858	0.875
300	0.903	0.898	0.851	0.870
(j) CNN-BiLSTM [10]				
k	10	20	30	40
40	0.907	0.891	0.860	0.880
50	0.897	0.867	0.855	0.869
100	0.907	0.894	0.894	0.855
150	0.928	0.877	0.8809	0.863
200	0.910	0.883	0.857	0.879
250	0.903	0.861	0.842	0.855
300	0.891	0.891	0.866	0.870

In order to study virus specific characteristics [11], we utilize Frequency Distribution plots using NLTK[5] as shown in Fig. 2. The plots show steep slopes and sharp corners indicating particular k-mers in each of the viruses which determine a signature which can help in reducing computational resources as shown in Table 1, where the FastText models trained using the most frequent k-mers gave better results as compared to using the entire set of unique k-mers as the vocabulary for training.

t-Distributed Stochastic Neighbor Embedding is a dimensionality reduction technique which is well suited for visualizing high-dimensional data. Figure 3 shows the efficiency of both the embedding methods for our dataset.

---

[5] https://www.nltk.org/.

Table 3. Classification results with dna2vec embeddings [8]

(a) Accuracy for CNN [9] and CNN-BiLSTM [10]

Classifier	Accuracy
CNN [9]	0.645
CNN-BiLSTM [10]	0.642

(b) Cross-validation scores for Machine Learning Classifiers

Classifier	Cross validation score
Naive Bayes	0.624
Decision Tree	0.819
SVC (Linear Kernel)	0.484
SVC (Polynomial Kernel)	0.656
SVC (RBF Kernel)	0.422
Linear SVC	0.783
Random Forest	0.819
AdaBoost	0.659

5.2   Conclusion and Future Scope

In this paper, we aim at analysing three viruses belonging to the same biological family viz. MERS, SARS and SARS-CoV-2 by experimenting with various NLP based embedding techniques and evaluate them using Machine Learning and Deep Learning classifiers. For computing sequences, we explore nine different methods - two using FastText and seven using GloVe, by generating k-mers with 'k' values- 10,20,30,40. We compare our results with the dna2vec [8] embeddings. We infer that FastText embeddings outperform other methods for all 'k' values and also are efficient in handling OOV, essential for dealing with virus mutations. For all embeddings, the classification accuracy seems to gradually increase with an increase in k-mer length as against the traditional use of shorter k-mers [8]. Using FastText, we observe that embeddings trained using the most frequent k-mers show significantly better results as compared to those trained with all the unique k-mers in the dataset. This paves a way to research regarding signature k-mers belonging to each virus which might contribute towards efficient computation.

In this paper, the sequence embeddings have been obtained by averaging the k-mer vectors. This implies that the sequence embedding is purely based on the contents of its k-mers . Considering the high classification performance, it could be hypothesized that any virus and its mutations fundamentally have an unchanging composition irrespective of the structural arrangement, which can be distinguished clearly from other virus sequences belonging to the same family or subgenus. This can be proven with adequate domain knowledge from biologists, and also through further experimentation using more samples. Thus, as a future scope, we wish to experiment with a larger dataset which incorporates more number of mutations.

References

1. Koyama, T., Platt, D., Parida, L.: Variants of the SARS-CoV-2 genomes. Bull. World Health Organ. **98**, 495–504 (2020)
2. Malik, Y.A.: Properties of coronavirus and SARS-CoV-2. Malays. J. Pathol. **42**(1), 3–11 (2020). PMID: 32342926
3. Lan, T.C.T., et al.: Structure of the full SARS-CoV-2 RNA genome in infected cells
4. Junior, J.A.C.N., Santos, A.M., Quintans-Júnior, L.J., Walker, C.I.B., Borges, L.P., Serafini, M.R.: SARS, MERS and SARS-CoV-2 (COVID-19) treatment: a patent review. Expert Opin. Ther. Pat. **30**(8), 567–579 (2020)
5. Li, Q., et al.: The impact of mutations in SARS-CoV-2 spike on viral infectivity and antigenicity. Cell **182**(5), 1284–1294 (2020)
6. Bojanowski, P., Grave, E., Joulin, A., Mikolov, T.: Enriching word vectors with subword information. Trans. Assoc. Comput. Linguist. **5**, 135–146 (2017)
7. NCBI Virus. https://www.ncbi.nlm.nih.gov/labs/virus/vssi
8. Ng, P.: dna2vec: consistent vector representations of variable-length k-mers. arXiv preprint arXiv:1701.06279 (2017)
9. Lopez Rincon, A., et al.: Accurate identification of SARS-CoV-2 from viral genome sequences using deep learning. bioRxiv (2020)
10. Zhang, J., Chen, Q., Liu, B.: DeepDRBP-2L: a new genome annotation predictor for identifying DNA binding proteins and RNA binding proteins using convolutional neural network and long short-term memory. In: IEEE/ACM Transactions on Computational Biology and Bioinformatics
11. Jha, P.K., Vijay, A., Halu, A., Uchida, S., Aikawa, M.: Gene expression profiling reveals the shared and distinct transcriptional signatures in human lung epithelial cells infected with SARS-CoV-2, MERS-CoV, or SARS-CoV: potential implications in cardiovascular complications of COVID-19. Front Cardiovasc Med. 7, 623012 (2021). Accessed 15 Jan 2021
12. Wang, L., Zhou, J., Wang, Q., Wang, Y., Kang, C.: Rapid design and development of CRISPR-Cas13a targeting SARS-CoV-2 spike protein. Theranostics. **11**(2), 649–664 (2021). Accessed 1 Jan 2021
13. Heo, L., Feig, M.: Modeling of severe acute respiratory syndrome coronavirus 2 (SARS-CoV-2) proteins by machine learning and physics-based refinement (2020)
14. Mikolov, T., Corrado, G., Chen, K., Dean, J.: Efficient estimation of word representations in vector space. In: Proceedings of the International Conference on Learning Representations (ICLR 2013), pp. 1–12 (2013)
15. Kwan, H.K., Arniker, S.B.: Numerical representation of DNA sequences, pp. 307–310 (2009). https://doi.org/10.1109/EIT.2009.5189632
16. Lopez-Rincon, A., et al.: Classification and specific primer design for accurate detection of SARS-CoV-2 using deep learning. Sci. Rep. **11**(1), 1–11 (2021)
17. Ballesio, F., et al.: Determining a novel feature-space for SARS-CoV-2 sequence data (2020)
18. Asgari, E., Mofrad, M.R.: Continuous distributed representation of biological sequences for deep proteomics and genomics, PLoS One **10**, e0141287 (2015)

19. Kimothi, D., et al.: Distributed representations for biological sequence analysis. ArXiv abs/1608.05949 (2016). n. Pag
20. Le, N.Q.K., Yapp, E.K.Y., Nagasundaram, N., Yeh, H.Y.: Classifying promoters by interpreting the hidden information of DNA sequences via deep learning and combination of continuous FastText N-grams. Front. Bioeng. Biotechnol. **7**, 305 (2019)

Use of Attitude Verbs as a Device to Encode States of Mind: Navigating Them Compositionally in Linguistic Discourse

Arka Banerjee[✉][iD]

Jadavpur University, Kolkata, India
banerjeesoumyo29@gmail.com

Abstract. This paper offers a computational distinction between two types of states of mind of an agent in linguistic discourse where attitude verbs such as *think, believe, etc.* are involved. These attitude verbs can be claimed to contribute to two different kinds of meaning layers, depending on the expectancy of the discourse. These two meaning layers are *at-issue layer* and *conventional implicature (CI) layer*. Here we provide a compositional distinction between these two kinds of contributions of the mental state denoting verbs, using the logic of CIs ($\mathcal{L}_{CI}$).

Keywords: States of mind · Attitude verbs · At-issue content/MPU · CI · QUD · $\mathcal{L}_{CI}$ · At-issue application · CI application

1 Introduction

Linguistic discourse can encode agent's states of mind in various means or strategies. These strategies of encoding them may include use of speaker-oriented adverbs (*e.g. fortunately, luckily, etc.*), use of expressives (*e.g. damn, freaking, etc.*), incorporation of attitude verbs like *think, believe, doubt,* and so on. Let us exemplify this with the following line of an imaginary accident report:

(1)　*Unfortunately, Suresh, the bus driver, succumbed on the way to the hospital.*

The underlined word in (1) indicates writer's state of mind towards the accident. That is, writer finds the death of the bus driver very unfortunate. In the same way, we can get hold of the speaker's mental state when he utters the sentence - *I won't go to the damn house*; we get a sense where the speaker does not like the house he is talking about. Analogously, another linguistic device of encoding speaker's mental state is by incorporating attitude verbs in discourse. Echoing [5], speakers can use language to report on the mental state of some individual - what he thinks, believes, *etc.* Thus, the use of attitude verbs is of immense linguistic importance in shedding light on an agent's state of mind in any form of discourse.

© Springer Nature Switzerland AG 2022
R. Chbeir et al. (Eds.): MIKE 2021, LNAI 13119, pp. 174–181, 2022.
https://doi.org/10.1007/978-3-031-21517-9_17

In this paper, we address the issue of using attitude verbs as a medium to denote agent's mental states in different ways. Following [8]'s idea, attitude verbs can have two different kinds of status in a linguistic discourse, namely (a) main point of the utterance (henceforth, MPU), and (b) non-MPU. This paper diagnoses the latter as a type of conventional implicature (henceforth, CI) and distinguishes it from the former in a compositional way, making use of [6]'s logic of CIs. Though the formalism for distinguishing them reflects the principle of compositionality [4], the task of distinction reminisces the Gricean framework [2] which provides us the instrumental machinery for conceptualizing a detailed meaning graph of any linguistic discourse.

The next section discusses the MPU and non-MPU status associated with attitude verbs, with respect to question-answer discourse. In Sect. 3, we explain that the non-MPU interpretation is related to the meaning of CI-level. In Sect. 4, we compositionally differentiate these two types of attitudinal status within [6]'s logic of CI ($\mathcal{L}_{CI}$). Lastly, Sect. 5 concludes the paper.

2   Two Kinds of Status of Attitude Verbs in Discourse: MPU and Non-MPU

There has been a multitude of discussions on the status of attitude verbs in linguistic literature. It is espoused in [1,12] that when a declarative sentence is subordinated under an attitude verb like *think*, *believe* and so on, the whole/entire clause becomes the asserted proposition. For example, in the sentence - *I think John is not in town*, the more important proposition is the whole sentence (*i.e.*, the main clause), not the subordinate or embedded clause - *John is not in town*. But, there are other cases where the embedded clause gets more preference in a discourse [11]. In such cases, [11] calls this type of uses of the embedding verb *parenthetical*. The following (Q)uestion-(A)nswer discourse instantiates a parenthetical use of the attitude verb *think*:

(2) Q: Why did not John attend the invitation last night?

A: I think John is not in town.

[8] assumes that the content which constitutes the answer to the question asked becomes the main point of the utterance or MPU. Therefore, the embedded clause in the above answer, *i.e., John is not in town* gets the main-point status in (2) because the proposition - *John is not in town* constitutes the probable answer to the question. And, the attitude verb *think* becomes parenthetical or non-MPU. Putting it differently, the function of *think* in (2) turns out to be evidential - it only amounts to considering how much reliable the source of information is. As opposed to it, the MPU status of *think* can be noted in the following Q-A discourse:

(3) Q: Why did not Sue invite John in her house party last night?

A: Sue thought John is not in town.

In a scenario where it turned out later that John was in fact in town in that night, the answer in (3) instantiates an MPU status of *think* because Sue did not invite John based on her thought that he is not in town. Thus the whole clause *i.e., Sue thought John is not in town*, but not the embedded clause alone, becomes the potential answer to the above question. In other words, the whole clause gets the asserted status in (3), with the attitude verb being the main point of the utterance.

Thus, we have seen that states of mind of an agent can be associated with two different types of meaning layer, though not simultaneously, based on the expectancy of the question in discourse. The next section will discuss more on this issue, with a highlight on what constitutes primary contents in a linguistic discourse and what does not.

3   Parenthetical States of Mind are CIs

In the previous section, we have shown that the contributions of *think* are not similar in (2) and (3). In the former, the attitude verb remains parenthetical or optional while in the latter, it constitutes a part of the main content. In this section, we establish arguments in favor of equating the use of parenthetical (non-MPU) attitude verbs with CI-entailment.

Prior to this argument, we should revert back to [2] who establishes the famous distinction between 'what is favoured' *vs.* 'what is conventionally implicated' in a conversational discourse. See the following excerpt ([2], pp. 44–45):

(4)   'In some cases the conventional meaning of the words used will determine what is implicated, besides helping to determine what is said. If I say (smugly), *He is an Englishman; he is, therefore, brave*, I have certainly committed myself, by virtue of the meaning of my words, to its being the case that his being brave is a consequence of (follows from) his being an Englishman. But while I have said that he is an Englishman and said that he is brave, I do not want to say that I have SAID (in the favored sense) that it follows from his being an Englishman that he is brave, though I have certainly indicated, and so implicated, that this is so. I do not want to say that my utterance of this sentence would be, STRICTLY SPEAKING, false should the consequence in question fail to hold.'

Following this insight, [6] suggests two major subtypes under *commitments* - (a) at-issue entailments, and (b) CI-entailments. The former corresponds to what is called main point of the utterance (MPU), or what is favored in Gricean term. In above passage in (4), it is implicated, but not said (in favored sense), that the man being brave follows from his being an Englishman. Let us first dive into the matter of what it really means for a linguistic item to have an at-issue entailment.

[9] define at-issueness with respect to the concept of Question Under Discussion (QUD) which goes back to [7]. QUDs are semantic questions, explicit or implicit in discourse. For example, take (2) where the question is explicit. A

QUD-stack may be implicit in a scenario in which interlocutors did not find John in the invitation. Let us repeat the Q-A sequence of (2) to illustrate how QUD acts on the moves (both questions/setup moves and assertions/payoff moves) of the interlocutors. See the following:

(5) m_1 : Why did not John attend the invitation last night? (setup move)
Ans(m_1)[1]: I think John is not in town. (payoff move)

The ordering function $<$[2] in the information structure of the concerned discourse yields the following order: $\langle m_1, \text{Ans}(m_1) \rangle$ where m_1 is uttered before Ans(m_1). Now, the function of QUD is as in what follows:

(6) $\text{QUD}(m_1) = \varnothing$[3]; $\text{QUD}(\text{Ans}(m_1)) = \langle m_1 \rangle$

Informally, QUD is giving us the set of all 'as-yet unanswered but answerable, accepted questions' at the time of any move. See [7] for the detailed formalization of QUD, which we will not discuss here. Another crucial notion that [7] offers is that of *relevance* which says that 'a move m is relevant to the question under discussion q, *i.e.*, to $last(\text{QUD}(m))$[4], iff m either introduces a partial answer to q (m is an assertion) or is part of a strategy to answer q (m is a question)'. With this notion of relevance, [9] define *at-issueness* as in what follows: (a) A proposition p is at-issue iff the speaker intends to address the QUD via ?p, (b) An intention to address the QUD via ?p is **felicitous** only if: (i) ?p is relevant to the QUD, and (ii) the speaker can reasonably expect the addressee to recognize this intention. In their definition, ?p means *whether p*; if the proposition is *I think John is not in town*, then ?*I think John is not in town* is tantamount to uttering *Do you think John is not in town?*. Using this definition of at-issueness, we can now exhibit that the main clause of the answer in (2) does not form the at-issue content in the discourse. Let us see how.

A clause can contain more than one utterance [10]. Following this intuition, the answer in (2) can express two utterances - (a) *Do you think John is not in town?*, and (b) *Is John not in town?*. Among these two, the second one becomes relevant to the question Q in (2), *i.e.*, to $last(\text{QUD}(\text{Ans}(m_1)))$ or m_1 in (5), because the question - *Is John not in town?* has an answer which contextually entails a partial answer to the Q, *viz.* *Why did not John attend the invitation last night?* (à la [9]). On the contrary, the answer to *Do you think John is not in town?* does not contextually entail either a partial or a complete answer to the Q in concern. Hence, the embedded proposition - *John is not in town* becomes the at-issue element in (2) and the attitude verb *think* gets a parenthetical (non-MPU) status there. We argue that the use of this attitude verb in this case exemplifies a commitment of CI-level, instead of an at-issue one. As mentioned in ([6], p.

---

[1] Ans(m) translates into 'the answer of the move m'.
[2] $<$ is the precedence relation, a total order on set of moves M:
$m_i < m_k$ iff m_i is made/uttered before m_k in discourse D. (See [7].)
[3] The empty set indicates that there are no moves prior to m_1.
[4] As per [7], for any move m, $last(\text{QUD}(m))$ or the immediate question-under-discussion at the time of m denotes the last element of the ordered set $\text{QUD}(m)$.

11), properties of CIs are the following: (a) CIs are part of the conventional meaning of words, (b) CIs are commitments, and thus give rise to entailments, (c) These commitments are made by *the speaker of the utterance* 'by virtue of the meaning of' the words he chooses, (d) CIs are logically and compositionally independent of what is '*said* (in the favored sense)', *i.e.*, independent of the at-issue entailments. Following this insight, the use of *think* in (2) instantiates a speaker-commitment which is independent of the embedded at-issue content. In other words, it is an implicature, though not *conversational* in nature because the implicated content cannot be denied after uttering it.

As opposed to that in (2), we argue that the entire answer clause forms the at-issue content in (3), because the answer to the question *Did Sue think John is not in town?* contextually entails the partial answer to the immediate question-under-discussion at the time of the payoff move in (3). Thus, the use of *think* here gets the status of main point of the utterance, being part of the at-issue entailment. It no longer remains parenthetical in (3).

Thus, it is noted that the mental states of an agent can have both at-issue and CI status, depending on the expectancy of the QUDs involved in the discourse. Now we are in a position to formalize the distinction between these two status in a computational way. In course of doing that, we will embrace the logic of conventional implicatures ($\mathcal{L}_{CI}$) as proposed by [6]. With reference to the basic framework of $\mathcal{L}_{CI}$, the next section discusses the formal distinction between the two status of attitude verbs in QUD discourse.

4   A Compositional Profile

In $\mathcal{L}_{CI}$, e^a, t^a, and s^a are basic at-issue types, while e^c, t^c, and s^c are basic CI types. It is assumed that e is the type of entities, t and s are types of truth values and worlds respectively. Another convention that needs to be mentioned here is - if '$\alpha : \tau$' is written, it would mean that the expression α is of type τ. These syntactic types serve as categories for lambda expressions. [6] summarizes two rules regarding how the comabinatorics work: (a) At-issue meanings apply to at-issue meanings to produce at-issue meanings, (b) CI meanings apply to at-issue meanings to produce CI meanings. To capture these two rules, the following two clauses are offered in the syntax of $\mathcal{L}_{CI}$: (a) If σ and τ are at-issue types for $\mathcal{L}_{CI}$, then $\langle \sigma, \tau \rangle$ is an at-issue type, (b) If σ is an at-issue type for $\mathcal{L}_{CI}$ and τ is a CI type for $\mathcal{L}_{CI}$, then $\langle \sigma, \tau \rangle$ is a CI type. It is always prohibited that at-issue meanings apply to CI meanings, because this possibility goes against the view that CIs are peripheral, nonintrusive elements. In the system of $\mathcal{L}_{CI}$, the following two kinds of *functional applications, i.e., at-issue application* (7a) and *CI application* (7b) are used for sake of the combinatorics:

(7) a. $[_{\alpha(\beta)\,:\,\tau^a} \quad \alpha : \langle \sigma^a, \tau^a \rangle \bullet \gamma : \rho^c \quad \beta : \sigma^a \bullet \delta : \upsilon^c]$

b. $[_{\beta\,:\,\sigma^a \,\bullet\, \alpha(\beta)\,:\,\tau^c} \quad \alpha : \langle \sigma^a, \tau^c \rangle \bullet \gamma : \rho^c \quad \beta : \sigma^a \bullet \delta : \upsilon^c]$

The basic mechanism for functional application is same as is proposed in [3]. The metalogical device '$\bullet$' is used to separate at-issue elements from CI elements.

With this mechanism of $\mathcal{L}_{CI}$, we will now provide computational analyses using lambda calculus for the MPU and non-MPU status of the states of mind associated with the attitude verbs in discourse. In (2) where the attitude verb is parenthetically used, it can be argued that the entire clause, namely *I think John is not in town* becomes the CI element, whereas the embedded clause - *John is not in town* gets the at-issue or MPU status. We assume that the matrix CI clause of the form 'I think p' is adjoined to the at-issue clause, 'John ($\mathbf{j}$) is not in town' yielding the following structure:

(8)

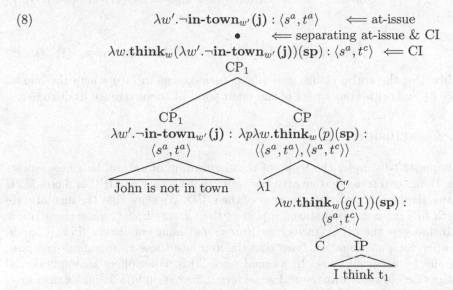

$$\lambda w'.\neg\textbf{in-town}_{w'}(\mathbf{j}) : \langle s^a, t^a\rangle \quad \Longleftarrow \text{at-issue}$$

$$\bullet \quad \Longleftarrow \text{separating at-issue \& CI}$$

$$\lambda w.\textbf{think}_w(\lambda w'.\neg\textbf{in-town}_{w'}(\mathbf{j}))(\mathbf{sp}) : \langle s^a, t^c\rangle \quad \Longleftarrow \text{CI}$$

The standard assumption of intensional logic is followed here in viewing propositions as $\langle s, t\rangle$-type elements, *i.e.*, functions from possible worlds to truth values. We are assuming that the embedded proposition in the adjoined clause is uninterpreted and moves to the left periphery of the same clause, binding the trace t_1 that is valued by the assignment function g. This adjoined CP becomes a CI element, because speaker's ($\mathbf{sp}$) thinking about the at-issue proposition is secondary with respect to the QUD discourse in (2). The adjoined clause composes with the at-issue CP$_1$ by CI application that is schematized in (7b). The resultant shows that the at-issue clause is independent of the matrix clause which is not main point of utterance in the concerned discourse.

As opposed to that, the attitude verb *think* is not parenthetical in QUD discourse (3). We argue it has the interpretation like below:

(9) $think \rightsquigarrow \lambda p_{\langle s^a, t^a\rangle} \lambda x_{e^a} \lambda w_{s^a}. \underbrace{\textbf{think}_w(p)(x)}_{\text{at-issue assertion}} : \langle\langle s^a, t^a\rangle, \langle e^a, \langle s^a, t^a\rangle\rangle\rangle$[5]

In (9), it is important to note that the assertion part - 'x thinks p in w' is not a CI material, rather an at-issue item. This is apparent from the type designation

---

[5] The symbol '$\rightsquigarrow$' stands for 'translates into'.

of it, *i.e.*, the at-issue t^a type. In this case, the compositional computation for the answer in (3) will look like (10), in which at-issue application is used:

(10)

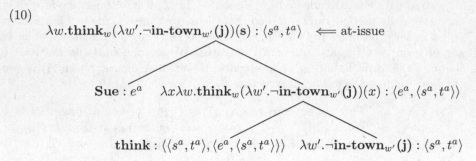

$$\lambda w.\mathbf{think}_w(\lambda w'.\neg\mathbf{in\text{-}town}_{w'}(\mathbf{j}))(\mathbf{s}) : \langle s^a, t^a \rangle \quad \Longleftarrow \text{at-issue}$$

$$\mathbf{Sue} : e^a \quad \lambda x \lambda w.\mathbf{think}_w(\lambda w'.\neg\mathbf{in\text{-}town}_{w'}(\mathbf{j}))(x) : \langle e^a, \langle s^a, t^a \rangle \rangle$$

$$\mathbf{think} : \langle\langle s^a, t^a \rangle, \langle e^a, \langle s^a, t^a \rangle\rangle\rangle \quad \lambda w'.\neg\mathbf{in\text{-}town}_{w'}(\mathbf{j}) : \langle s^a, t^a \rangle$$

Unlike (8), the entire clause gets the at-issue status in (10) where the mental state of Sue (**s**) becomes part of the main point of the utterance in discourse.

5   Conclusion

This paper falls under the scope of the engineering of natural language semantics. Here, we discuss that an attitude verb can get either an MPU or a non-MPU status, depending on the expectancy of the QUD. We show that the mental state which has the non-MPU status is part of the CI entailment, while the attitude verb that gets the MPU status constitutes an at-issue entailment. In this paper, we offer computational analyses that differentiate these two readings compositionally by using [6]'s $\mathcal{L}_{\mathrm{CI}}$. In a broad sense, this paper offers a computational distinction between primary and secondary information in a linguistic discourse.

Acknowledgements. Thanks to Utpal Lahiri for his insights. Errors are mine.

References

1. Abbott, B.: Presuppositions as nonassertions. J. Pragmat. **32**(10), 1419–1437 (2000)
2. Grice, P.H.: Logic and conversation. In: Cole, P., Morgan, J. (eds.) Syntax and Semantics, iii: Speech Acts, pp. 43–58. Academic Press, New York (1975)
3. Heim, I., Kratzer, A.: Semantics in Generative Grammar. Blackwell, Oxford (1998)
4. Partee, B.H.: Compositionality. In: Landman, F., Veltman, F. (eds.) Varieties of Formal Semantics, pp. 281–311. Foris, Dordrecht (1984)
5. Pearson, H.: Attitude verbs. In: Matthewson, L., Meier, C., Rullmann, H., Zimmermann, T.E. (eds.) Companion to Semantics. Wiley, Oxford (2020)
6. Potts, C.: The Logic of Conventional Implicatures. OUP, New York (2005)
7. Roberts, C.: Information structure in discourse: toward an integrated formal theory of pragmatics. Semant. Pragmatics **5**(6), 1–69 (2012)
8. Simons, M.: Observations on embedding verbs, evidentiality, and presupposition. Lingua **117**(6), 1034–1056 (2007)
9. Simons, M., Tonhauser, J., Beaver, D., Roberts, C.: What projects and why. In: Proceedings of SALT, vol. 20, pp. 309–327 (2010)

10. Tonhauser, J.: Diagnosing (not-)at-issue content. In: Proceedings of SULA, vol. 6, pp. 239–254 (2012)
11. Urmson, J.O.: Parenthetical verbs. Mind **61**(244), 480–496 (1952)
12. Wilson, D., Sperber, D.: Ordered entailments: an alternative to presuppositional theories. In: Oh, C.K., Dinneen, D.A. (eds.) Syntax and Semantics 11: Presupposition, pp. 299–323. Academic Press, New York (1979)

CAI: Complex Ontology Alignments Using Lexical Indexation

Houda Akremi[1(✉)] and Sami Zghal[1,2(✉)]

[1] Faculty of Sciences of Tunis, LIPAH-LR11ES14, University of Tunis El Manar,
2092 Tunis, Tunisia
houda.akremi@gmail.com
[2] Faculté des Sciences Juridiques, Économiques et de Gestion de Jendouba,
Université de Jendouba, Campus Universitaire, 8189 Jendouba, Tunisia
sami.zghal@fsjegj.rnu.tn

Abstract. Ontology matching is the process of automatically determining the semantic equivalences between the concepts of two ontologies. However, simple ontology alignments are a domain largely studied. It link one entity of source ontology to one entity of target ontology. In this paper, we present a complex ontology alignment system CAI. In contrast to most existing tools, CAI is a matching system that can deal with complex and simple ontology alignments.

Keywords: Ontology · Similarity · Indexation · Complex alignment · Simple alignment

1 Introduction

The use of ontologies, as well as reference representations in various fields (such as medicine, agronomy, etc.) has led to a heterogeneity of information. Most of the proposed ontology alignment methods are conventional approaches [2]. Indeed, each concept of the first ontology can be mapped to a concept of the second ontology [8]. Thus, complex ontology alignment methods make it possible to map a concept of the first ontology to several concepts of the second ontology [7]. The research related to simple alignment has been well studied [9]. In addition, due to the complexity of the alignments between ontologies, only identifying traditional simple 1-to-1 alignment is not enough to fulfill the growing high demand of most of these applications [11]. Many automatised ontology matching approaches proposed in the literature deal with simple alignment generation. They link identifiers of entities to each other, they find correspondences like $(O_1 : Paper, O_2 : Paper, \equiv)$ which declares equivalent the Paper concepts of the ontologies O_1 and O_2. The need for complex correspondences has been identified in various domains and applications. The main challenge is to create new alignment systems and methods to uncover complex relations in real-world use cases. In this context, a new complex alignment method of ontologies is proposed. Despite the variety of automatic matching approaches, most of them aim

© Springer Nature Switzerland AG 2022
R. Chbeir et al. (Eds.): MIKE 2021, LNAI 13119, pp. 182–187, 2022.
https://doi.org/10.1007/978-3-031-21517-9_18

at fully aligning two ontologies, the output alignment aims at fully covering the common scope of the two ontologies. However, a user may not need as much coverage as he or she may be interested by only a part of the ontology scope.

The paper is organised as follows. Section 2 gives a background on complex alignments and discuses the related work. Section 3 focuses on the proposed approach in alignments ontologies. Section 4 summarizes an evaluation of our approach of complex alignments with a complex alignment dataset. Finally, Sect. 5 concludes this work and discusses its perspectives.

2   Complex Alignments

Ontology matching is the process of generating an alignment A between two ontologies: a source ontology O_1 and a target ontology O_2 [8]. A is a set of correspondences, denoted $A_{O_1 \to O_2}$. Each correspondence is a triplet (e_{O_1}, e_{O_2}, r). e_{O_1} and e_{O_2} are the members of the correspondence (classes, object properties, data properties, instances, values, etc.) of respectively O_1 and O_2 or constructions of these entities using constructors or transformation functions. r is a relation, equivalence ($\equiv$), a subsumption ($\sqsubseteq, \sqsupseteq$), or a disjointedness ($\perp$) between the correspondences e_{O_1} and e_{O_2}. Thereafter, ontologies alignment contains two types of correspondences simple and complex.

Correspondence is simple [5], both e_{O_1} and e_{O_2} are atomic entities: one single entity is matched with another single entity. However, correspondence is complex, if at least one of e_{O_1} or e_{O_2} involves a constructor or a transformation function. The complex ontology matching is the process of generating a complex alignment between ontologies. In fact, complex alignment generation is more difficult than simple alignment generation. Indeed, the matching space between two ontologies is the set of all the possible correspondences between them. In simple alignment generation, the matching space is limited to the product of the source entities and the target entities. If O_1 contains m entities and O_2 contains n entities, the matching space is $m \times n$. In complex alignment generation, the matching space is larger as the complex expressions can include any number of entities, constructors and transformation functions.

Complex alignments are more expressive than simple alignments. The specificities of the latter rely on their output and their process. Many works in the literature have focused on complex alignments [1,3,4,6].

The AROA system (Association Rule-based Ontology Alignment) [4] proposes a set of matching conditions to detect correspondence patterns. The conditions are based on the labels of the ontology entities, the structures of these ontologies and the compatibility of the data-types of data-properties. The matching conditions to detect these patterns are an input to the matching algorithm. The user can add new matching conditions to detect other patterns.

The KAOM (Knowledge-Aware Ontology Matching) [3] generates transformation function correspondences and logical relation correspondences. KAOM implements different matching strategies: one for detecting transformation function correspondences, the others for logical relation correspondences. Here we present its transformation function correspondence detection approach, as it uses an atomic pattern. A Kullback-Leibler divergence measure on the data values is used to define the coefficient of the linear transformation.

The CANARD system [6] deals with binary CQAs (Competency Questions for Alignment), *i.e.*, CQAs whose expected answers are pairs of instances or literal values. The difference between CANARD 2018 and CANARD 2019 is mainly based on case of unary CQAs because the two instances of the pair answer are matched instead of one.

AgreementMakerLight (AML) [1] is an ontology matching system inspired on Agreement-Maker and drawing on its design principles, but with an added focus on scalability to tackle large ontology matching problems. AML is primarily based on lexical matching algorithms, but also includes structural algorithms for both matching and filtering, as well as its own logical repair algorithm. It makes use of external biomedical ontologies and the WordNet as sources of background knowledge.

The specificities of the complex alignment generation approaches rely on their output and their process. These are the two axes of the proposed classification. Most approaches require external output such as matched instances or a simple alignment. Overall, the approaches look for a way to reduce the matching space between two ontologies. To do so, they rely on structures, common instances, external resources such as web query interfaces or a combinations of those. However, all of them intend to cover the full scope of the source and target ontologies. As such, in this paper, we propose a new complex alignments ontology approach CAI (Complex ontology alignments using lexical Indexation) that focuses on two aspects: lexical indexation and Competency Questions for Alignment.

3   Description of CAI

The CAI method consists of four phases: parsing phase, indexing phase, candidate identification phase and alignment phase. Indeed, it starts with a parsing phase that allows to model the input ontologies in a format that can be used for the rest of the process. The second phase, continues with an indexing phase over the considered ontologies. Then these indexes are queried to supply the candidate alignments in the third phase. Finally, CAI generates the alignment file.

3.1   Parsing Phase

The parsing phase (or pretreatment phase) is crucial for complex matching ontologies. It is performed using the OWL API[1]. In fact, the parsing phase transforms the considered ontologies represented initially as two OWL files in an adequate format for the rest of the treatments. In CAI approach, the purpose is to substitute all the existing information in both OWL files. Indeed, each

---

[1] http://owlapi.sourceforge.net/.

entity is represented by all its properties. The parsing phase begins by loading two ontologies to align. In other words, this phase is accompanied by a linguistic pretreatment phase. Linguistic pretreatment involves eliminating special characters, filtering empty words, and lemmatizing. The linguistic pretreatment prepare the ontologies to be aligned during the indexing phase. The resulting informative format, has a considerable impact on the similarity computation results thereafter. As a result, the parsing phase gets couples formed by the entity and its associated label.

3.2   Indexing Phase

The indexing phase consists of reducing search space through the use of an effective search strategy on indexes. To ensure this phase CAI index both ontologies (O_1 and O_2) and generate two index vectors. In fact, the indexing process makes it possible to create at the ontology level a set of indexes which make it easier to search. This search is done by integrating the different generated indexes. Thus, indexing exploits the vector model. This model makes it possible to represent the ontology in the form of a vector. Thus, each element of the index vector consists of an ontological concept accompanied by its descriptors (label, comment, sub-concepts, relation, axioms, etc.). The resulting of the indexing phase is a set of indexes for the two ontologies. The set of indexes is used in the candidate identification phase.

3.3   Candidate Identification Phase

The new complex ontology alignment method proposed the discovery of the complex correspondences based Competency Questions for Alignment. In our case, we use CQA to find the entities in common between the indexes. Indeed, the CQA represent the knowledge needs of a user and define the scope of the alignment [5]. In CAI, the queries role is to find the entities in common between the indexes. In this stage we deal with binary n-ary CQA whose answers are a tuple of instances or more. The result of this phase is a set of documents sorted by relevance. Thereafter, our system continues with counting similarity score.

3.4   Phase of Alignment

To align the candidates, the proposed process takes as input the two ontologies O_1 and O_2, it carries out the extraction of the different ontological entities that compose them. Then, it creates the indexes corresponding to each of the ontologies to apply the correspondence thereafter.

4   Evaluations

The Ontology Alignment Evaluation Initiative (OAEI), proposes a new track of ontology alignment evaluation report Complex alignments[2]. In our case the

---

[2] http://oaei.ontologymatching.org/2018/complex/index.html.

dataset Conference is used. This dataset is based on the OntoFarm dataset [10] used in the Conference track of the OAEI campaigns. It is composed of 16 ontologies on the conference organisation domain and simple reference alignments between 7 of them. We consider 3 out of the 7 ontologies from the reference alignments (cmt, conference and ekaw), resulting in 3 alignment pairs. For this first evaluation, only equivalence correspondences will be evaluated and the confidence of the correspondenes will not be taken into account. Table 1 shows the number of classes, and object properties, and data properties in cmt, conference and ekaw Ontologies.

Table 1. Characteristics of cmt, conference and ekaw

Name	Classes	Datatype properties	Object properties
Ekaw	74	0	33
Cmt	36	10	49
Conference	14	2	15

In fact, CAI presents results of precision, recall and F-measure encouragement. As for the conference test set, it consists of heterogeneous ontologies which represent the organization of conferences and which are used in real applications. CAI finishes with 100.00% precision, 56.41% recall and 72.10% F-measure for the simple (1 : 1) correspondences. Thus, for complex (1 : n) correspondences CAI presents 66.38% precision, 100.00% recall and 79.80% F-measure. From the performance comparison, only CAI and CANARD [6] can generate almost correct complex alignment (Table 2.) For instance, our system CAI can produce correct entities that should be involved in a complex alignment, but it doesn't output the correct relationship. Based on these situations, we will investigate the incorrect alignments and improve the algorithm to find the relationship and entities as accurate as possible.

Table 2. Performance comparison on conference benchmark

Evaluation metrics	CAI	AML	AROA	CANARD
Precision	**66.38**%	50.00%	0.86%	89.00%
Recall	**100.00**%	23.00%	46.00%	39.00%
F-measure	**79.80**%	32.00%	60.00%	54.00%

5   Conclusion

In this paper, we have proposed a new ontology complex alignment method. We first proposed indexing to create the corresponding indexes for each ontology.

Then, a process of identifying candidates is applied. Experiments have shown that our alignment system achieves the best performance. Future work includes the exploration of a more improved ontology complex alignment model.

References

1. Faria, D., Pesquita, C., Santos, E., Cruz, I.F., Couto, F.M.: Agreement maker light results for OAEI 2013. In: Proceedings of the 8th International Conference on Ontology Matching (OM 2013), vol. 1111, pp. 101–108, Aachen, DEU (2013). https://ceur-ws.org/
2. Jain, P., Hitzler, P., Sheth, A.P., Verma, K., Yeh, P.Z.: Ontology alignment for linked open data. In: Patel-Schneider, P.F., et al. (eds.) ISWC 2010. LNCS, vol. 6496, pp. 402–417. Springer, Heidelberg (2010). https://doi.org/10.1007/978-3-642-17746-0_26
3. Jiang, S., Lowd, D., Kafle, S., Dou, D.: Ontology matching with knowledge rules. In: Hameurlain, A., Küng, J., Wagner, R., Chen, Q. (eds.) Transactions on Large-Scale Data- and Knowledge-Centered Systems XXVIII. LNCS, vol. 9940, pp. 75–95. Springer, Heidelberg (2016). https://doi.org/10.1007/978-3-662-53455-7_4
4. Ritze, D., Meilicke, C., Šváb-Zamazal, O., Stuckenschmidt, H.: A pattern-based ontology matching approach for detecting complex correspondences. In: 4th ISWC Workshop on Ontology Matching, vol. 551, pp. 25–36 (2009)
5. Thiéblin, E., Haemmerlé, O., dos Santos, C.T.: CANARD complex matching system: results of the 2018 OAEI evaluation campaign. In: International Workshop on Ontology Matching Co-located with the 17th ISCW (OM@ISWC 2018), pp. 138–143, Monterey, US (2018)
6. Thiéblin, E., Haemmerlé, O., Trojahn, C.: Complex matching based on competency questions for alignment: a first sketch. In: 13th International Workshop on Ontology Matching Co-located with the 17th International Semantic Web Conference (OM@ISWC 2019), pp. 66–70, Monterey, US (2019)
7. Todorov, K., Hudelot, C., Popescu, A., Geibel, P.: Fuzzy ontology alignment using background knowledge. Int. J. Uncertain. Fuzziness Knowlege-Based Syst. **22**(1), 75–112 (2014)
8. Xue, X., Yao, X.: Interactive ontology matching based on partial reference alignment. Appl. Soft Comput. **72**, 355–370 (2018)
9. Xue, X., Zhang, J.: Matching large-scale biomedical ontologies with central concept based partitioning algorithm and adaptive compact evolutionary algorithm. Appl. Soft Comput. **106**, 107343 (2021)
10. Zamazal, O., Svátek, V.: The ten-year OntoFarm and its fertilization within the onto-sphere. Journal of Web Semantics **43**, 46–53 (2017)
11. Zhou, L.: A journey from simple to complex alignment on real-world ontologies. In: DC@ISWC (2018)

Long Short-Term Memory Recurrent Neural Network for Automatic Recognition of Spoken English Digits

Jane Oruh and Serestina Viriri[✉]

School of Mathematics, Statistics and Computer Science,
University of KwaZulu-Natal, Durban 4000, South Africa
viriris@ukzn.ac.za

Abstract. Automatic recognition of isolated spoken digits is one of the utmost demanding tasks in Automatic Speech Recognition (ASR) because of its complexity. Recently, deep learning approaches have been deployed for this task, and this has outperformed the traditional machine learning approaches such as ANN. Particularly; deep learning methods such as Long Short-Term Memory (LSTM) has achieved improved performance in ASR. LSTM method is however restricted to processing continuous input streams that are not segmented into sub-sequences. In this work, an enhanced deep learning LSTM RNN model is proposed to resolve this shortcoming. In the proposed model, a Recurrent Neural Network (RNN) is incorporated as a "forget gate" to the memory block to allow resetting of the cell states at the beginning of sub-sequences. The architecture of the proposed model will in turn modify the standard architecture of the LSTM networks, to be able to make effective use of the model parameters while addressing the computational efficiency problems of large networks for large vocabulary speech recognition. The result of the experiment shows an overall accuracy of 99.36% on the well-established public benchmark spoken English digits dataset through a deep supervised learning approach, and the Back Propagation-Through Time (BPTT) method of Adam optimization algorithm for updating error weight parameter.

Keywords: Deep supervised learning · Recurrent neural network · Automatic speech recognition · Spoken English digits

1 Introduction

Speech comprises sequences of uttered sounds, also known as phonemes. Speech is used to transmit information from one speaker to the other. When the signal from a speech is converted into a meaningful message or text, it is called Automatic Speech Recognition (ASR) [15]. Recognition of isolated spoken digits has proven to be a challenging task in the area of ASR due to its complexity. In recent times, deep learning has been deployed for ASR in embedded systems [11], speech recognition systems [6], and has outperformed the traditional machine learning approaches such as Artificial Neural Networks (ANN).

© Springer Nature Switzerland AG 2022
R. Chbeir et al. (Eds.): MIKE 2021, LNAI 13119, pp. 188–198, 2022.
https://doi.org/10.1007/978-3-031-21517-9_19

Although ANNs can categorize small acoustic-phonetic units, such as the separate phonemes, they cannot model long-term dependencies in the acoustic signals [19]. However, the Deep Neural Networks (DNNs) provides restricted temporal modeling for the acoustic frames. They cannot deal with data that have longer-term dependencies. Feed-forward neural networks can be expanded towards effective classification. To achieve this, it will require feeding the signals back into the network from previous time steps. Such networks having recurrent interconnections are called Recurrent Neural Networks (RNN) [28,29].

The connections in RNNs are cyclic, which makes it a dynamic mechanism for modeling sequence data [22]. Thus, RNNs use a dynamically contextual window against a static fixed-sized window over sequences. Unfortunately, RNNs are hard to train using the gradient-based Back Propagation Through Time (BPTT) [21], and are not likely to demonstrate the full power of recurrent models. This is because of the well-known vanishing gradient and exploding gradient problems [2].

One way to improve the training of RNNs is to make use of an optimization algorithm with higher-order approximations [13]. Although, it is normally at the cost of remarkable increased computational costs, which makes the approach unattractive for language modeling that requires an enormous size of training data [25]. Hochreiter and Schmidhuber [10], proposed the Long Short-Term Memory (LSTM) architecture, as a solution to resolving the challenge.

A modified LSTM RNN model has been proposed in this work, to perform sequence prediction that will make use of deep supervised learning on the benchmark spoken English digits dataset. The effectiveness of the model will be estimated in respect of training and validation accuracy, and the result will be compared with other works that used the LSTM technique for various speech recognition tasks. In addition, the model's classification performance will be evaluated to obtain the average score for precision, recall, and f1-score using a confusion matrix. The choice of LSTM RNN is based on the fact that LSTM consists of a standard RNN built up with "memory units", that specializes in transferring long-term information, also with a set of "gating" units that allows memory units to carefully interrelate with the normal RNN hidden state [13].

Several works have been done using LSTM RNN before this time. LSTM has achieved virtually all thrilling results that are based on RNNs. Thus, it has turned out to be the centre of deep learning in the ASR system [31]. LSTMs has been used largely on speech recognition tasks because of their powerful learning ability [4,6,9,20,22,26,27], but this is the first time LSTM RNN will be used on the spoken English digits speech recognition dataset.

The contributions of this research can be summarized as follows;

1. A modified LSTM that permits input streams that are not divided into sequences to be continuously processed at the beginning of sub-sequences.
2. The research work made use of Adam optimization algorithm to update error weight parameters through a BPTT approach.

3. The proposed system was evaluated for the first time on the well-established public benchmark spoken English digit dataset, and the overall accuracy of 99.36% was achieved.

The rest of this work is organized as follows. Section 2 reviews the previous work in the field of ASR using LSTM. Section 3 describe the proposed methods and techniques. Section 4 discusses the experimental setup, training and results for the proposed model, and presents a comparative analysis. Finally, Sect. 5 presents the conclusion.

2   Related Work

Graves et al.'s [6] work shows that end-to-end training methods like CTC (Connectionist Temporal Classification) can be used to train RNNs for sequence labelling tasks on the TIMIT corpus. Merging these methods with LSTM RNN architecture will likely yield state-of-the-art results. This paper has used the standard LSTM RNN training method to obtain 99.36% accuracy for the sequence prediction speech recognition task.

Sak et al. [22] work, was found to have introduced the first implementation of LSTM networks on the Google voice search speech recognition task. Their proposed model architecture made more improved use of model parameters while training acoustic models. The model trained and compared LSTM, RNN, and DNN models at various numbers of parameters and configurations. The result shows that the LSTM model was the fastest to converge and performs best when applied on a moderately small-sized framework.

In [4], the authors proposed an LSTM RNN in a hybrid acoustic modelling structure for robust speech recognition in an environment affected by noise and reverberation. The experiment was conducted on the database of the 2nd CHiME medium-vocabulary recognition track. The authors compared state prediction networks and networks predicting phonemes using LSTM networks. The result of their experiment shows that with the use of LSTMs in a hybrid or double stream system, the state prediction network is superior to the network prediction phonemes.

He and Droppo [9] has proposed a generalized LSTM known as (G)LSTM-DNN. The strength of the proposed model was first analyzed using normal 80-hour LVCSR task AMI, and then applied to the 2000-hour Switchboard data set. The result of their experiment has shown that the proposed (G)LSTM-DNN performs better with more layers, and achieved the relative word error rate reduction of 8.2% on the 2000-hour Switchboard data set. One issue as was discussed in their work is that the model's performance comes at the cost of a huge parameter number, it would be noteworthy to find a system that will save the parameter while maintaining its modeling power.

Tachioka and Ishii [26], proposed LSTM RNN for Bandwidth Extension (BWE) on the TIMIT phoneme recognition task. The proposed LSTM RNN-based BWE was compared with the standard GMM-based BWE. The outcome of the experiment has shown LSTM RNN-based BWE more powerful than the

GMM-based BWE. Also, they added that for ASR purposes, it is better to predict Mel Frequency Cepstrum Coefficients (MFCC) features directly than to predict Mel-cepstrum features. The model used in this work has used MFCC features for its prediction.

The authors had proposed LSTM-RNN for deep sentence embedding [16]. Here, RNN is used to accept each word in a sentence sequentially and then mapped alongside the contextual information into a latent space in a recurrent form. Furthermore, the LSTM cells were incorporated into the RNN model (LSTM-RNN) to address the weakness of RNN in learning long-term memory. As a result of the non-availability of labeled data, user click-through data was used and the model was trained in a weak supervised form. The proposed LSTM RNN used in this work for the sequence prediction on spoken English digit data, however, was trained using a strong deep supervised network that helped in obtaining optimal accuracy.

One of the RNN models, namely, gated recurrent units (GRUs) was revised, and a simpler architecture that proved to be more advantageous was proposed in [20] for automatic speech recognition across various tasks, features, conditions, and paradigms. The experiment was conducted using TIMIT, the DIRHA-English, CHiME, and TED-talk speech recognition corpus, in various subsections. The proposed method has outperformed the standard GRU in terms of recognition and computational performance and has significantly reduced per-epoch training time by 30% as against the standard GRU.

WAZIR and CHUAH [27] have used the standard LSTM networks to train the large vocabulary spoken Arabic digit speech recognition task. Their model showed the result for an overall accuracy of 94.00% for model training, and 69.00% for the model testing. When the standard LSTM was being implemented on the spoken English digits speech recognition task, the overall accuracy of 99.36% was obtained for model training as demonstrated in this work.

3   Methods and Techniques

3.1   The Proposed LSTM Architecture

The main structure of the LSTM consists of unique segments known as "memory blocks" in the hidden layer. The first type of LSTM block comprises cells, input and output gates. The standard structure of the LSTM has a limitation, which was addressed for the first time in [5] through the establishment of a "forget gate" that will empower LSTM to adjust its state. The "forget gate" resets the cell variable leading to the 'forgetting' of the stored input c_t, whereas the "input" and "output" gates manage the reading of inputs from the feature vector, x_t, and writing of output to h_t, respectively [10].

The gates regulate the action of the memory block whereas the "forget gate" weighs the information inside the cells, such that anytime previous information becomes unimportant for some cells, it will reset the state of the different cells. Forget gates also enables continual prediction [12], by making cells forget their previous state, thereby restricting biases in prediction.

Figure 1 shows the proposed LSTM RNN Memory blocks.

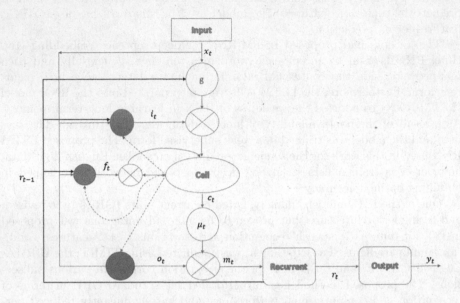

Fig. 1. Proposed enhanced LSTM RNN architecture.

The LSTM uses a forward pass approach to solve the following equations:

$$i_t = \sigma(W_{ix}X_t + W_{ih}h_{t-1} + b_i) \tag{1}$$

$$f_t = \sigma(W_{fx}X_t + W_{fh}h_{t-1} + b_f) \tag{2}$$

$$c_t = f_t \odot c_{t-1} + i_t \odot tanh(W_{cx}X_t + W_{ch}h_{t-1} + b_c) \tag{3}$$

$$o_t = \sigma(W_{ox}X_t + W_{oh}h_{t-1} + b_o) \tag{4}$$

$$h_t = o_t \odot tanh(c_t) \ [24]. \tag{5}$$

From the equations above, W variables are the weight matrices; W_{ih}, W_{fh}, W_{oh} and b variables are the bias vectors, with b_i as the input gate bias vector. The operation $\odot$ denotes the element-wise vector product. The σ, g and h are point-wise non-linear activation functions, and i, f, o and c are the input gate, the "forget gate", the output gate and the cell activation vectors, respectively. The entire features of the LSTM network architecture can be trained with the sigmoid(ϕ) and $tanh$ activation functions.

The proposed model integrates RNN as a "forget gate" to the memory block to permit cell states to be reset at the beginning of sub-sequences. The memory blocks use its memory cells to store the network's temporal state, and distinctive multiplicative units known as gates to control information flow. The proposed model architecture makes effective use of the model parameters by modifying the standard LSTM architecture.

Supervised learning is a learning technique that makes use of labeled data. For a supervised deep learning technique, the setting is made up of a set of inputs with complementary output $(x_t, y_t) \sim p$. For instance, if for an input x_t, the smart agent predicts $\hat{y} = (x_t)$, then the agent will get a loss value $l=(y_t, \hat{y}_t)$. After successful training, the agent will adjust the network parameters repeatedly to get an improved approximation of the output, similar to the deep supervised approach used in this work [1].

Algorithm 1 represents the algorithm of the proposed Model

Algorithm 1. Proposed Speech Recognition Model for the LSTM-RNN Network

1: **procedure** ENHANCED LSTM RNN PROCEDURE(X, Y)
2:     Input Speech (Speech X)
3:     Extract Feature Map;
4:     LSTM processing;
5:     RNN processing - Cell's state memory resetting;
6:     LSTM processing;
7:     Model training
8:     Generate Prediction (Map$Y$);
9:     Perform Optimal Estimation using Adam Optimization;
10:     Output Recognized Speech
11: **end procedure**

---

4 Experiments

4.1 Dataset

The dataset used for this work is made up of isolated spoken digits. It is a tar file consisting of 15 different speakers (male and female). Each speaker utters a digit 16 times, which leads to $15 * 16 = 240$ instances for each digit. The phrases are English numbers: 0–9. This gives us a total of 2400 different audio files in a wav format for training the proposed system.

The dataset is a publicly available dataset under Pannous, a collaboration working on improving speech recognition [17], which is from the librosa library [14]. The speech data was downloaded using the MFCC batch generator. The file consists of a group of wav files that are in batches alongside its related labels. The audio dataset was pre-processed using the functionality of the librosa library, one of Python's several libraries dedicated to analyzing sounds.

The dataset was split into training and validation sets. 10% of the dataset was used for validation, whereas the remaining 90% was used for training. The training step output contained validation accuracy and loss as shown in Table 1 since the validation set was introduced as part of the model fit function during training.

The proposed LSTM RNN network structure comprised of four network layers; an input layer, LSTM (dropout) layer, the fully connected layer and a regression layer. The model is being trained using a deep learning library known as TFLearn.

4.2 Proposed Model Training

Learning rate and training iterations could affect the accuracy and training time of the proposed model. Therefore, both parameters were adjusted with different values for optimal performance. Given that the learning rate should be considered as the most crucial hyperparameter, it might therefore be necessary to understand how to adjust it properly to achieve a positive outcome [7]. The learning rate regulates the speed of the network's weight updates. The initial learning rate of the model is set to 10^{-3}.

Following is the training iterations, which was adjusted to the initial value of 1000 iterations. The training iterations were used in multiplying the epoch size to get the training steps. The training steps, with 10 epochs and the batch size of 64/64, ranged from 10000 to the 20000 training steps. High accuracies were achieved when the training steps were increased.

To reduce LSTM's total loss on a set of training sequences, Adam optimization algorithm was used to improve the parameter of each network weight to the weight parameter using the BPTT method [3, 21]. The BPTT method, used for learning the weight matrices of an RNN unravels the network on time and disseminates error signals backwards through time. The major challenge with the BPTT method is the problem of vanishing gradients. But, by making use of LSTM cells, this difficulty is being overcome to a great extent [30].

Cross-entropy loss that made use of SoftMax activation function was used in training the networks. At the initial learning rate of 10^{-3}, it was observed that the model trained fast, but the accuracy was dropping due to overfitting. By adjusting the learning rate to 10^{-4}, the model trained slowly, with the increase in network accuracy.

4.3 Proposed Model Training Outputs

The computational graphs of the model's output are visualized through a TensorBoard. Time-dependent scalar statistics that vary over time and variations in accuracy performance are visualized in Figs. 2 and 3, respectively.

4.4 Results and Discussions

The result of the model's training has shown that good hyperparameter, such as the learning rate, helps to manage a large set of experiments for hyperparameter tuning. This shows that increasing the learning rate will lead to network training fast, whereas reducing the learning rate will lead to accurate prediction of the

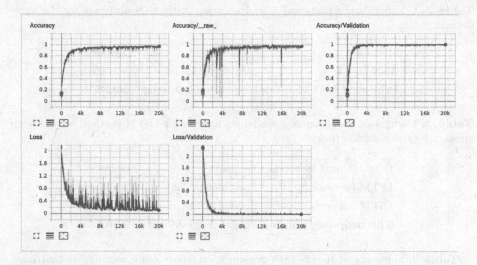

Fig. 2. Model's accuracy and loss for 10^{-3} learning rate @2000 training iterations.

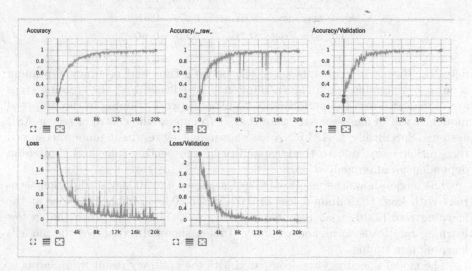

Fig. 3. Model's accuracy and loss for 10^{-4} learning rate @2000 training iterations.

Table 1. The result for learning rates tuning and its corresponding accuracy.

Training iterations	Learning rate	Loss	Accuracy	Val loss	Val acc
1000	0.001	0.49798	0.9197	0.02684	0.9844
1000	0.0001	0.31974	0.9333	0.01019	1.0000
2000	0.001	0.10913	0.9665	0.00189	1.0000
2000	0.0001	0.02656	0.9936	0.00130	1.0000

Table 2. Comparing the proposed model accuracy with ANN accuracy for the same dataset.

Model	Dataset	Result
ANN [23]	Spoken English digit	82.00%
Proposed LSTM model	Spoken English digit	**99.36%**

Table 3. Comparing the proposed model accuracy with LSTM method accuracy on spoken digit recognition tasks.

Deep learning methods	Dataset	Accuracy
LSTM network [8]	Free spoken digit	95.00%
LSTM network [27]	Spoken Arabic digit	94.00%
The proposed model	Spoken English digit	**99.36%**

Table 4. Summary of the model's classification report using a confusion matrix.

Average score	Precision	Recall	F1-score
Macro average	0.67	0.67	0.61
Weighted average	**0.70**	**0.60**	**0.60**

network. Hence, it represents trade-offs between time and accuracy. Optimum accuracy is possible when the learning rate is reduced, and the training steps are increased. From the performance result of network training in the proposed model, it will be necessary to state that RNNs are at the centre of recent ASR systems. Specifically, LSTM RNN has shown exciting results in numerous speech recognition tasks, due to their capability in representing long and short-term dependencies in sequences [18].

The model shows the result of 99.36% accuracy and 100.00% validation accuracy with least the minimal loss of 0.02656 for 2000 training iterations at the learning rate of 10^{-4}, as represented in Table 1. This is to prove that a low learning rate leads to higher accuracy. Table 1 shows the result of the model's learning rate tuning.

The model's accuracy was compared with the accuracy result from Sarma et al.'s [23] work, which used ANN for spoken digit recognition on the same dataset. The result shows an overall accuracy of 82.00%, as against 99.36% accuracy achieved by using LSTM RNN in this work. This justifies the earlier statement that ANNs models cannot model long-term dependencies, a clear limitation of the ANNs. The comparison of the two techniques was summarized in Table 2. A further comparison was made with other works that used the LSTM technique on spoken digit recognition tasks, as shown in Table 3.

To further investigate the classification performance of the model, a confusion matrix was used to generate the classification report on the average score for precision, recall, and f1-score of the proposed model as shown in Table 4.

5 Conclusion

For the first time, this work shows that LSTM RNN models can rapidly train large vocabulary speech recognition of the spoken English digits, through a deep supervised learning approach that makes effective use of the model parameter tuning. The model proposed in this work predicted real numbers of the spoken English digit data. By implementing the proposed method, 99.36% accuracy was achieved, which outperformed some other deep learning methods. Thus, the proposed model has addressed the computational competence that is required for training large networks.

References

1. Alom, M.Z., et al.: A state-of-the-art survey on deep learning theory and architectures. Electronics 8(3), 292 (2019)
2. Bengio, Y., Simard, P., Frasconi, P.: Learning long-term dependencies with gradient descent is difficult. IEEE Trans. Neural Netw. 5(2), 157–166 (1994)
3. Boden, M.: A guide to recurrent neural networks and backpropagation. Dallas Proj. 2(2), 1–10 (2002)
4. Geiger, J.T., Zhang, Z., Weninger, F., Schuller, B., Rigoll, G.: Robust speech recognition using long short-term memory recurrent neural networks for hybrid acoustic modelling. In: Fifteenth Annual Conference of the International Speech Communication Association (2014)
5. Gers, F.A., Schmidhuber, J., Cummins, F.: Learning to forget: continual prediction with LSTM. Neural Comput. 12(10), 2451–2471 (1999)
6. Graves, A., Mohamed, A.R., Hinton, G.: Speech recognition with deep recurrent neural networks. In: 2013 IEEE International Conference on Acoustics, Speech and Signal Processing, pp. 6645–6649. IEEE (2013)
7. Greff, K., Srivastava, R.K., Koutník, J., Steunebrink, B., Schmidhuber, J.: LSTM: a search space odyssey. IEEE Trans. Neural Netw. Learn. Syst. 28, 2222–2232 (2017)
8. Mahalingam, H., Rajakumar, M.: Speech recognition using multiscale scattering of audio signals and long short-term memory of neural networks. Int. J. Adv. Comput. Sci. Cloud Comput. 7(2), 12–16 (2019)
9. He, T., Droppo, J.: Exploiting LSTM structure in deep neural networks for speech recognition. In: 2016 IEEE International Conference on Acoustics, Speech and Signal Processing (ICASSP), pp. 5445–5449. IEEE (2016)
10. Hochreiter, S., Schmidhuber, J.: Long short-term memory. Neural Comput. 9(8), 1735–1780 (1997)
11. Lin, S., et al.: FFT-based deep learning deployment in embedded systems. In: 2018 Design, Automation & Test in Europe Conference & Exhibition (DATE), pp. 1045–1050. IEEE (2018)
12. Lyu, Q., Zhu, J.: Revisit long short-term memory: an optimization perspective. In: Advances in Neural Information Processing Systems Workshop on Deep Learning and Representation Learning, pp. 1–9. Citeseer (2014)
13. Martens, J., Sutskever, I.: Learning recurrent neural networks with Hessian-free optimization. In: ICML (2011)
14. McFee, B., et al.: Librosa: v0.4.0. Zenodo, 2015. In: Proceedings of the 14th Python in Science Conference (SCIPY 2015) (2015)

15. Nasreen, P.N., Kumar, A.C., Nabeel, P.A.: Speech analysis for automatic speech recognition. In: Proceedings of International Conference on Computing, Communication and Science (2016)
16. Palangi, H., et al.: Deep sentence embedding using long short-term memory networks: analysis and application to information retrieval. IEEE/ACM Trans. Audio Speech Lang. Process. **24**(4), 694–707 (2016)
17. Pannous.Github: Pannous/tensorflow-speech-recognition. http://github.com/pannous/tensorflow-speech-recognition (2016). Accessed 3 May 2020
18. Parcollet, T., Morchid, M., Linarès, G., De Mori, R.: Bidirectional quaternion long short-term memory recurrent neural networks for speech recognition. In: ICASSP 2019 IEEE International Conference on Acoustics, Speech and Signal Processing (ICASSP), pp. 8519–8523. IEEE (2019)
19. Ramesh, K.V., Gahankari, S.: Hybrid artificial neural network and hidden Markov model (ANN/HMM) for speech and speaker recognition. Int. J. Comput. Appl. **975**, 8887 (2013)
20. Ravanelli, M., Brakel, P., Omologo, M., Bengio, Y.: Light gated recurrent units for speech recognition. IEEE Trans. Emerg. Top. Comput. Intell. **2**(2), 92–102 (2018)
21. Rumelhart, D.E., Hinton, G.E., Williams, R.J.: Learning representations by back-propagating errors. Nature **323**(6088), 533–536 (1986)
22. Sak, H., Senior, A., Beaufays, F.: Long short-term memory based recurrent neural network architectures for large vocabulary speech recognition. arXiv preprint arXiv:1402.1128 (2014)
23. Sarma, P., Sarmah, S., Bhuyan, M.P., Hore, K., Das, P.P.: Automatic spoken digit recognition using artificial neural network. Int. J. Sci. Technol. Res. **8**(12), 1400–1404 (2019)
24. Sennhauser, L., Berwick, R.C.: Evaluating the ability of LSTMs to learn context-free grammars. arXiv preprint arXiv:1811.02611 (2018)
25. Sundermeyer, M., Schlüter, R., Ney, H.: LSTM neural networks for language modeling. In: Thirteenth Annual Conference of the International Speech Communication Association (2012)
26. Tachioka, Y., Ishii, J.: Long short-term memory recurrent-neural-network-based bandwidth extension for automatic speech recognition. Acoust. Sci. Technol. **37**(6), 319–321 (2016)
27. Wazir, A.S.M.B.A., Chuah, J.H.: Spoken Arabic digits recognition using deep learning. In: 2019 IEEE International Conference on Automatic Control and Intelligent Systems (I2CACIS), pp. 339–344. IEEE (2019)
28. Werbos, P.J.: Backpropagation through time: what it does and how to do it. Proc. IEEE **78**(10), 1550–1560 (1990)
29. Williams, R.J., Zipser, D.: A learning algorithm for continually running fully recurrent neural networks. Neural Comput. **1**(2), 270–280 (1989)
30. Yu, D., Deng, L.: Recurrent neural networks and related models. In: Automatic Speech Recognition. SCT, pp. 237–266. Springer, London (2015). https://doi.org/10.1007/978-1-4471-5779-3_13
31. Yu, Y., Si, X., Hu, C., Zhang, J.: A review of recurrent neural networks: LSTM cells and network architectures. Neural Comput. **31**(7), 1235–1270 (2019)

Prediction of Smoking Addiction Among Youths Using Elastic Net and KNN: A Machine Learning Approach

Shreerudra Pratik[1] , Debasish Swapnesh Kumar Nayak[2(✉)] ,
Rajendra Prasath[3(✉)] , and Tripti Swarnkar[4(✉)]

[1] Department of Mathematics, Utkal University, Bhubaneswar 751004, India
[2] Department of Computer Science and Engineering, Institute of Technical Education and Research, Siksha 'O' Anusandhan Deemed to be University, Bhubaneswar 751030, India
swapnesh.nayak@gmail.com
[3] Computer Science and Engineering Group, Indian Institute of Information Technology Sri City, Chittoor 517646, India
rajendra.prasath@iiits.in
[4] Department of Computer Application, Institute of Technical Education and Research, Siksha 'O' Anusandhan Deemed to be University, Bhubaneswar 751030, India
triptiswarnakar@soa.ac.in

Abstract. In the current generation, it has been studied that smoking addiction among the youths is increased exponentially. Since there is a lot of awareness among the people about tobacco use but the youths are exposed to a broader spectrum of the different types of nicotine products like E-cigarettes, or hookah or water pipes, or conventional cigarettes, or dissolvable tobacco, and many more products. Our study aims are to identify some of the demographic factors like age, gender, sex, etc. to predict smoking addiction among the youths of our society. To predict e-cigarette addiction among youths, we considered the wings of Artificial Intelligence (AI) like Machine Learning (ML), and Deep Learning (DL) for better understanding and finding the relationship among the various features. During the data preprocessing we used the elastic net regression technique for feature selection and K-Nearest neighbor for making the accurate predictions. The Hybrid Prediction model was built by the Elastic net regression technique and KNN. It is observed that Elastic net finds a better selection of significant features. The outcome of the suggested pipeline provides high performance on the selection of significant features for the prediction model and provides better accuracy.

Keywords: E-cigarettes · Machine Learning (ML) · Deep Learning (DL) · Elastic net · K-Nearest Neighbor (KNN)

1 Introduction

In the modern world, despite knowing the harmful effects of using tobacco products, youths are getting attracted to it. Tobacco products are easily available substances in

R. Chbeir et al. (Eds.): MIKE 2021, LNAI 13119, pp. 199–209, 2022.
https://doi.org/10.1007/978-3-031-21517-9_20

the market. Different types of tobacco products that are available in the market are E-cigarettes, water pipes, hookah, dissolvable tobacco, snus, conventional cigarettes, chewing tobacco, etc. Among those products, E-cigarettes and conventional cigarettes are widely used by the youths. E-cigarettes are cell-operated gadgets that produce vaporized solutions to inhale. E-cigarettes are easily available devices that may look like a traditional cigarette and can be available in different types of flavors such as mint, chocolate, clove or spice, etc. [3, 6, 9, 10, 14, 17, 19]. In some aspects of life, it can be used for nicotine replacement therapy to quit smoking [7, 18]. Some of the youths are using e-cigarettes as a form of status and some of them are using this because it can be used anywhere and everywhere. But there is a concern that e-cigarettes aerosols contain some harmful chemicals which can become very harmful for the lungs and other respiratory problems [8]. An overview directed somewhere in the year of 2014 as well as 2018 appeared the pace of e-cigarette use in the youth of age between Eighteen to Twenty-four years to be 7.6%, this is deciphered as around one out of five teenagers presently utilizing e-cigarettes [11, 12]. There is expanding proof that e-cigarettes might be related to an expanded danger of oral sicknesses prediabetes, sorrow, asthma, (COPD) Constant Obstructive Pulmonary Disease, and respiratory symptoms [13]. E-cigarettes have been related to weed, non-endorsed drug use, and resulting cigarette smoking, which might be clarified by puzzling due to normal responsibility like shared hereditary weakness or natural components [12]. Users of E-cigarettes are increasing day by day among conventional cigarettes smokers, but e-cigarette use by never smokers is additionally rising. In 2016, 15% of all E-cigarette clients were sole e-cigarette clients and around 12 lakhs of them were not exactly or smaller than 25 years of age [15].

Like conventional cigarettes, E-cigarettes are also responsible for creating ultrafine particles which contain nicotine and this nicotine is delivered to the brain. E-cigarettes contain some substances like vegetable glycerol and propylene glycol, which are regularly added substances in e-fluids, can make extra unsafe synthetics—aldehydes when warmed [16]. With the progression in Machine Learning, inventive and principled variable determination procedures have been made available to applied analysts working with medical services information Utilizing these new improvements to character key determinants of the supplier patient conversations about smoking in a more comprehensive way may supplement ebb and flow research about the smoking end. The features related to e-cigarette can be distinguished by implementing various machine learning strategies.

In the era of Artificial Intelligence (AI), the techniques like Machine Learning, and Deep Learning are the subsets of it. Machine Learning is the space of computational science that spotlights on dissecting and deciphering examples and constructions in information to empower picking up, thinking, and dynamic outside of human cooperation. There has been an increment in the utilization of ML methods to medication and other examination regions however there is a lack of the utilization of Machine Learning methods in tobacco research. ML is a characteristic augmentation of customary factual methodologies that becomes expanding significantly as the measure of information increments and the dimensionality of the dataset increments. As the measure of factors to be viewed as increments, distinguishing every one of the factors related with a result and deciding the factors to be remembered for models turns out to be progressively hard to execute appropriately utilizing standard factual techniques. ML methods can be utilized to distinguish factors related to a result as the quantity of factors increment. ML procedures have been applied to review information to distinguish factors that are related to various mental and infection conditions. Deep Learning is also a subset of ML which has the capability of learning unsupervised data which are unstructured, it is also known as deep neural learning or deep neural network. With the expansion in the measure of huge datasets and the approach of designs preparing units (GPUs), algorithmic procedures and techniques have been consistently improved. Deep Learning was stretched out from classical machine learning by adding a few more profound constructions to models to consequently accomplish highlight extraction from crude information and has shown preferable execution over traditional machine learning for some classification and prediction models. As the dataset in the field of tobacco research is very large so, we used both machine learning and deep learning to understand the factors which are responsible for the addiction of smoking among the youths.

2 Related Work

Starting from LASSO otherwise known as the least absolute shrinkage and selection operator which is a regression analysis method that can be used for both variable selection and regularization can be used for increasing the accuracy of the prediction model. Through the concept of LASSO overfitting of data can be avoided when there is any bigger difference between the trained data and test data. LASSO regression is used when there is a larger dataset because it uses the L1 regularization technique which automatically performs feature selection.

LASSO can be used with other Machine Learning algorithms like BORUTA for the selection of the features linked with current e-cigarettes users by implementing a shrinking process where the value of least important features leads to zero by penalizing the coefficient of the regression features [1]. The method LASSO and BORUTA have found out the most identical features which can give a better understanding for the use of e-cigarettes among the young adults of U.S., But there is a concern that the research doesn't focus on the usage of E-cigarettes among the participants who had different types of disabilities, high-risk behaviors, and chronic conditions.

LASSO and Random Forest were used to build the prediction model to identify nicotine addiction among the youths [2]. To increase the Prediction accuracy, LASSO performs the regularization by selecting the important features and Random Forest has selected some of the subsets of features and constructing them in decision trees. During the feature selection, RF has found a fewer number of features than the LASSO, but some common features are found by both the Machine Learning Techniques.

3 Materials and Methods

To analyze the data and process it, we used the most important data mining technique which is Machine Learning. To identify the predictor features for determining the addiction of smoking among the youths, we used NYTS 2020 dataset. After getting the whole dataset we have used Elastic net and K-Nearest Neighbor (KNN) to identify and classify the smoking addiction predictor features and build the prediction model.

3.1 Dataset Used

To analyze data and process it, we have used Machine Learning, which works as feature selection from the "National Youth Tobacco Survey" 2020 dataset. Dataset is collected computerized and available on the public website with the URL ("https://www.cdc.gov/tobacco/data_statistics/surveys/nyts/data/index.html") [4]. In NYTS there were sixteen years of data starting from 1999 to 2020 in the zip format. Zip format contains dataset, codebook, Questionnaires, and methodology report in the format of.pdf, excel, MS-access. We have downloaded the excel format for our research. There were many features and tuples available in different years of data but we compared the last four-year NYTS dataset, we have found that in each successive year the features of the dataset are going on the increase, so we have used the 2020 data set for our further research. 2020 dataset is used for extracting the features that are responsible for the addiction to smoking among the youths (Fig. 1).

Fig. 1. This figure represents the character dataset where we have highlighted some of the identical fields.

3.2 Feature Selection

In our research work, we have used the Elastic Net regression model to identify the predictor features which are related to smoking addiction among youths. The redundant features having some coefficient can be made exactly to zero by the use of Elastic net. The coefficient values which are obtained by Elastic net are more stable and reliable so that we can use them for reducing the model complexity and to select quality-related features. The implementation of the Elastic Net to the NYTS 2020 dataset significantly reduces the computational time. During the implementation we found that the Elastic net feature selection model extracts the relevant features within 11.38 min with the system specification of Intel core i5-8300H 8th generation having 8 GB of RAM, 512 GB SSD, 4 GB NVIDIA GeForce GTX 1650, 2.30 GHZ with a 64-bit operating system.

$$\frac{\sum_{i=1}^{n}\left(y_i - x_i^j \hat{\beta}\right)^2}{2n} + \lambda\left(\frac{1-\alpha}{2} + \sum_{j=1}^{m} \hat{\beta}_j^2 + \alpha \sum_{j=1}^{m} \widehat{|\beta_j|}\right) \tag{1}$$

The Elastic net is the combination of the penalties of both LAASO and Ridge regression in order to get better features from the dataset. In the above Eq. 1 α is the mixing parameter between Ridge and LASSO where the α value of Ridge is 0 and the α value of LASSO is 1 and λ can be used as shrinkage parameter. Like LASSO Elastic net is also a variable selection method that does the work automatically, makes the shrinkage continuously, and can be able to select the groups of corresponding features [5]. The process of shrinkage in the Elastic net is done by combining the L_1 (norm penalty) of LASSO and L_2 (norm penalty) of the ridge. The function of L_1 is to produce a scattered model by decreasing some regression coefficients to zero and the function of L_2 is to balance the L_1 regularization path, encouraging the associated effects and removing the limitations on the selected features as shown in Eq. 1 [20].

3.3 Classification

After getting the structured data in tabular format, we have used K-nearest neighbor (KNN) for classifying the data in such a way that we can be able to get a high-performance predictive model which has better accuracy for determining the smoking addiction among the youths.

In the process of extracting the significant features, despite having K-NN or K-nearest neighbor is one of the successful supervised and non-parametric Machine Learning algorithms. It uses forecast dependent on the closeness of test guides to all accessible information. All the more explicitly, K-NN orders protest by essentially noticing k closest neighbors and along these lines translating a class mark that is pre-predominantly present in k picked environment. Considering the negative outcomes that might result from a deficient determination of k boundary, it is suggested that the number of neighbors ought to be odd, near the square base of the number of cases. Nonetheless, this heuristic ought not to be viewed as an overall arrangement and a fitting decision for all issues.

3.4 Flow Chart of Implemented Model

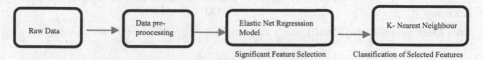

Fig. 2. Work-flow diagram of the model

The above block diagram represents the pipeline of the proposed model where we implemented the Elastic Net regression model for the feature selection and K-nearest neighbor for the classifying of the data (Fig. 2). In the above pipeline we have used the NYTS dataset for our research work which we got from NYTS website. After getting the data, preprocessing of the data have been done by eradicating the null values from the columns that were having more than 30%. Then the preprocessed data have been implemented in our model to select the features associated with the smoking addiction. After getting the features we have used KNN for the classifying of the data and finding the accuracy of the model.

4 Result Discussion

We identified predictor features and built the smoking addiction prediction models using machine learning algorithms for youth e-cigarette. The model built by the Elastic Net regression technique and K-Nearest Neighbor (KNN) shows high predictive performance. By using Elastic net we found out the predicted features from the available dataset, and then the classification and accuracy were measured by using the KNN. The model developed by the machine learning algorithm shows the robustness of the prediction performance.

The summary of the error values which were obtained from the dataset after the removal of null values. Different error values can be observed from the table based on Mean Absolute Error, Mean Squared Error, and Root Mean Squared Error on the test data set of Elastic Net Regression after applying the alpha value which is equal to 0.05, and L1_ratio which is equal to 0.5 as shown in Table 1 (Table 2).

Table 1. Mean Absolute Error, Mean Squared Error, and Root Mean Squared Error on test data by taking alpha value and L1_ratio as 0.05 and 0.5 respectively in Elastic Net feature prediction model.

Elastic Net	
Error Details	Values
Mean Absolute Error (MAE) on test data of E-Net Regression	0.425371175
Mean Squared Error (MSE) on test data of E-Net Regression	0.254893102
Root Mean Squared Error (RMSE) on test data of E-Net Regression	0.504869391

Table 2. Model accuracy Comparison of Elastic Net with LASSO

Parameters details	LASSO	Elastic-Net
Accuracy	0.6370	0.92
Root Mean Square Error (RMSE)	0.7509	0.5048

It is observed that the accuracy obtained in this work is significantly better than the work done so far [2]. It is also found that the result is relevant and the features extracted have responsible for increasing the addiction of youths in various smoking products like E-cigarettes, cigarettes, snus, hookah-waterpipe etc.

All the coefficient estimated values of the features that are selected by the Elastic net regression model are shown in Table 3, a sum of 26 features were selected by the model where some of them were in positive values and some of them were negative values. All the features extracted from the dataset were used for finding the accuracy of the prediction model. Here in the Elastic regression model, we used the alpha value and L1 ratio where alpha value is used to decide the amount of penalty that we want for the model, and the L1 ratio is the mixing parameter of the Elastic net i.e. $0 < = L1\_ratio <$

= 1. If the L1 ratio is zero then the penalty would be L2 (L2 penalty) which is nothing but Ridge regression and if the l1 ratio is one then the penalty would be L1 (L1 penalty) which is LASSO regression, so we have taken the value of L1_ratio between zero and one which means that the penalty would be the combination of L1 and L2 i.e. Elastic net regression technique.

It is observed that in our model The parameter setting is carried out by three different ways, taking the parameter value as 0 can be regarded as Ridge regression, taking the parameter value as 0.5 can be regarded Elastic net, and taking the parameter value 1 as LASSO. The result obtained from these three parameters setting methods is clearly indicate that the prediction value obtained by taking the parameter 0 and 1 is worst. While the prediction value obtained with parameter value 0.5 is significant with high accuracy.

The selected features from the dataset, which resembles that these features are responsible for the smoking addiction among the youths, Elastic net regression model selects 26 features to build the prediction model as shown in Table 3. The annotation of table data features with actual questions used during the survey is also shown.

Table 3. Represents the features which were selected by using Elastic net regression in order to determine the smoking addiction among the youths.

Sl no.	Features	Actual question	Estimated coefficient value
1	Q3	What is your grade?	0.00913412
2	Q9	In the last 30 days how many days you have consumed E-cigarettes?	0.00045497
3	Q17	Do you have the curiosity of using E-cigarettes?	0.01629883
4	Q31	Do you have curiosity about smoking cigarettes?	0.01572248
5	Q34	Are you going to smoke cigarette, if your friend offers you?	0.00088179
6	Q37	In the last 30 days how many days you have consumed Cigars/Cigarillos/Little Cigars?	0.00055838
7	Q52	Do you have curiosity about using tobacco in a hookah/water pipe?	0.00657235
8	Q58	The Number of days you have used nicotine product(s) in past 30 days?	0.00091843
9	Q76	Does anyone ask you about using e-cigarettes while visiting medical staff in the last twelve months?	0.00778932

(continued)

Table 3. (*continued*)

Sl no.	Features	Actual question	Estimated coefficient value
10	Q77	Does anyone advise you not to use e-cigarettes while visiting medical staff in the last twelve months?	0.01427373
11	Q78	Have you seen the warning label on the cigarette packet in the past 30 days??	0.00031264
12	Q85	Which is more or less or equal addictive than cigarettes (Cigars/Cigarillos/Little Cigars)?	0.0003782
13	Q86	Using the different types of nicotine products on some days can harm people?	0.01037961
14	Q88	Using E-cigarettes on some days can harm people?	0.01029183
15	Q89	E-cigarettes are less, more, or equally addictive than cigarettes?	0.00327961
16	Q90	Can hookah/waterpipe harm people, if they use it some days?	0.02282322
17	Q91	Which is more or less or equal addictive than cigarettes (hookah/water pipe)?	0.00592098
18	Q92	Contrasted with a run-of-the-mill Cigarette, could you imagine a cigarette publicized as low nicotine would be?	0.00750704
19	Q94	Excluding the fume from E-cigarettes, have you felt that absorbing smoke from others' cigarettes/other tobacco items causes …?	0.00365199
20	Q96	Out of ten students, number of students smoke cigarettes of the same grade?	0.00568515
21	Q100	At the point when you go to any shop, how frequently do you see advertisements/advancements for cigarettes/other tobacco items?	0.0110697
22	Q104	At the point when you go to any shop, how regularly do you see advertisements/advancements for e-cigarettes?	0.01141172

(*continued*)

Table 3. (*continued*)

Sl no.	Features	Actual question	Estimated coefficient value
23	Q105	At the point when you stare at the TV or real-time features or go out to see the films, how regularly do you see advertisements or advancements for e-cigarettes?	0.00130425
24	Q107	In the last seven days, how many times has someone used nicotine items in your presence?	0.01002481
25	Q109	How long did you absorb the smoke from somebody who was using some nicotine items in an indoor public spot in the last 30 days?	0.00092262
26	Q115	Which language do you speak at home other than English?	0.07815156

The features which were selected by the Elastic net regression model are shown in Fig. 3, after applying different alpha and L1_ratio we got 26 selected features from the dataset. The X-axis represents the number of features that were involved in the analysis and in Y-axis we have different coefficient values of the features. Some features have positive coefficient values and some of them have negative coefficient values. Different colors of lines that emerged from the X-axis are considered as the selected features for the analysis. The above selected features are very close to the previous work. However, the features extracted from our model is more accurate and it is found that these features are responsible for causing the addiction [2].

Fig. 3. The coefficient values of all the features

Overall performance of the model based on accuracy, precision, F1- score, and Recall of the multiclass classification of the prediction model, the accuracy of the model is 92% which is quite good. The performance of the model is measured by the confusion matrix of K-nearest Neighbor by putting the value of k = 1. The overall data of the model is represented in Table 4.

Table 4. Represents the classification as well as the performance of the model based on some parameters.

Classification	Accuracy	precision	F1-Score	Recall
Multiclass	0.92	0.93	0.95	0.96

5   Critical Analysis and Future Work

The predicting features identified by both Machine Learning and Deep Learning algorithms affirmed numerous socio-ecological determinants. It can be observed that, peer pressure and different types of brands of tobacco products as important predicting features. Studies show there are a few factors firmly related to youth smoking addiction, including their age, gender, grade, race, staying with regular tobacco users, English not being the main language at home, and having a handicap or emotional wellness condition. The results show that Elastic net's 26 predicting features include grade, curiosity, English is not becoming the main language at home, etc.

We had the option to fabricate AI-based nicotine fixation expectation models for the center and high schoolers utilizing freely accessible overview information. Although the review showed a decent return rate and the size of information was enormous, there were still limitations. It is observed from the literature that prediction of smoking addiction in youths is carried out by using various statistical and Machine Learning techniques. Statistical techniques cannot able to deal with the vast dataset and thus leads to computational complexity as well as result in error. There were some drawbacks which were emerged during the research work like during the training the dataset vast number of data are missing and it is also found that handling the preprocessing stage of the NYTS database is a very difficult and complex part because of the self-reported data. The overview of the NYTS database is cross-sectional, estimating reactions at a solitary preview on schedule, thus neglecting the improvement of smoking enslavement and reliance practices as they create over the long run. If any other Machine Learning algorithms were utilized, there may have been distinctive indicator factors showing diverse ideal execution. So we planned our next work is to focus on the more suitable preprocessing pipeline for the NYTS dataset.

References

1. Nkiruka, C., Atuegwu, C.O., Laubenbacher, R.C., Perez, M.F., Mortensen, E.M.: Factors associated with e-cigarette use in U.S. young adult never smokers of conventional cigarettes: a machine learning approach. Int. J. Environ. Res. Public Health **17**(19), 7271 (2021)
2. Choi, J., Jung, H.-J., Ferrell, A., Woo, S., Haddad, L.: Machine learning-based nicotine addiction prediction models for youth e-cigarette and waterpipe (Hookah) users. J. Clin. Med. **10**(5), 972 (2021)
3. Pariyadath, V., Stein, E.A., Ross, T.J.: Machine Learning classification of resting state functional connectivity predicts smoking status. Front. Hum. Neurosci. **8**, 425 (2014)
4. https://www.cdc.gov/tobacco/data_statistics/surveys/nyts/data/index.html. Last accessed on 4 June 2021

5. Ter Braak, C.J.F.: Regression by L1 regularization of smart contrasts and sums (ROSCAS) beats PLS and elastic net in latent variable model. J Chemomet. **23**(5)217–228 (2009)

6. Benowitz, N.L., Burbank, A.D.: Cardiovascular toxicity of nicotine: Implications for electronic cigarette use. Trends Cardiovasc. Med. **26**(6), 515–523 (2016)

7. Wang, D., Connock, M., Barton, P., Fry-Smith, A., Aveyard, P., Moore, D.: 'Cut down to quit' with nicotine replacement therapies in smoking cessation: a systematic review of effectiveness and economic analysis. Health Technol. Assess. 12(2), 2008

8. Kosmider, L., et al.: Carbonyl compounds in electronic cigarette vapors: Effects of nicotine solvent and battery output voltage. Nicotine Tob. Res. **16**(10),1319–1326 (2014)

9. Gentzke, A., et al.: Vital signs: tobacco product use among middle and high school students—United States, 2011–2018. Morb. Mortal. Wkly. Rep. **68**(6), 157–164 (2019)

10. Atuegwu, N.C., Perez, M.F., Oncken, C., Thacker, S., Mead, E.L., Mortensen, E.M.: Association between regular electronic nicotine product use and self-reported periodontal disease status: population assessment of tobacco and health survey. Int. J. Environ. Res. Public Health **16**(7), 1263 (2019)

11. McConnell, R., et al.: Electronic cigarette use and respiratory symptoms in adolescents. Am. J. Respir. Crit. Care. Med. **195**(8), 1043–1049 (2017)

12. Dutra, L.M., Glantz, S.A.: Electronic cigarettes and conventional cigarette use among U.S. adolescents: a cross-sectional study. JAMA Pediatr. **168**(7), 610–617 2014

13. Soneji, S., et al.: Association between initial use of e-cigarettes and subsequent cigarette smoking among adolescents and young adults: a systematic review and meta-analysis. JAMA Pediatr. **171**(8), 788–797 (2017)

14. Shahab, L., Beard, E., Brown, J.: Association of initial e-cigarette and other tobacco product use with subsequent cigarette smoking in adolescents: a cross-sectional, matched control study. Tob. Control **30** (2020)

15. Mirbolouk, M., et al.: E-cigarette use without a history of combustible cigarette smoking among U.S. adults: behavioral risk factor surveillance system, 2016. Ann. Intern. Med. **170**(1), 76–79 (2019)

16. Cheng, T.: Chemical evaluation of electronic cigarettes. Tob. Control **23**, ii11–ii17 (2014)

17. Sharma, A.: E-cigarettes compromise the gut barrier and trigger inflammation. Iscience **24**(2), 102035 (2021)

18. Wiemken, T.L., Kelley, R.R.: Machine learning in epidemiology and health outcomes research. Annu. Rev. Public Health **41**, 21–36 (2020)

19. Borland, R., Yong, H.H., O'Connor, R.J., Hyland, A., Thompson, M.E.: The reliability and predictive validity of the heaviness of smoking index and its two components: findings from the international tobacco control four country study. Nicotine Tob. Res. **12**(Suppl 1), S45–S50 (2010)

20. Reunanen, J.: Overfitting in making comparisons between variable selection methods. J. Mach. Learn. Res. **3**, 137–1382 (2003)

Novel Training Methods Based Artificial Neural Network for the Dynamic Prediction of the Consumed Energy

Arwa Ben Farhat[1]([⊠]) and Adnen Cherif[2]

[1] Electrical Engineering Department, ENICarthage, Charguia 2, Tunis, Tunisia
arwabenfarhat@gmail.com
[2] Physic Department, University of Tunis, El Manar, 1068 Tunis, Tunisia

Abstract. ANN have demonstrated best effectiveness and excellent scheduling capabilities in realizing many purposes like recognition, clustering, classification, management and even prediction. For this reason, we have used RBF based ANN for the dynamic prediction of load and PV production using many operations like forecasting, training and validation of the data accuracy. For the validation, the MAPE is calculated in function of the most three relevant input parameters, which are previous load and PV production measurements, seasonability and temperature or solar radiation data. This work has used real-time measurements of load and PV production for their comparison with the predicted load data using RBFNN algorithms to calculate the MAE and the MAPE, and to deduce the performance of the dynamic prediction algorithms and the accuracy of the forecasted data. In this context, this work has dealt with 2 key objectives. The first objective has dealt with the short-term dynamic prediction of load and PV production including forecasting and training operations. The 2nd objective is the calculation of MAE and MAPE via the comparison between the forecasted data and realtime measurements to evaluate the data accuracy and the performance of the dynamic prediction algorithms. By this way, the dynamic prediction algorithms were implemented, the predicted data were compared to the real measurements in the same time series and MAPE of the forecasted load data was calculated.

Keywords: Dynamic prediction algorithms · RBFNN · ISO · ErrCor · MAPE

1 Introduction

The predicting features identified by both Machine Learning and Deep Learning algorithms affirmed numerous socio-ecological determinants. It can be observed that, peer pressure and different types of brands of tobacco products as important predicting features. Studies show there are a few factors firmly related to youth smoking addiction, including their age, gender, grade, race, staying with regular tobacco users, English not being the main language at home, and having a handicap or emotional wellness condition. The results show that Elastic net's 26 predicting features include grade, curiosity, English is not becoming the main language at home, etc.

© Springer Nature Switzerland AG 2022
R. Chbeir et al. (Eds.): MIKE 2021, LNAI 13119, pp. 210–217, 2022.
https://doi.org/10.1007/978-3-031-21517-9_21

The use of smart meters is still unusable in many developing countries, despite its progression in developed countries [1, 2, 4]. For this reason, this work is aimed to forecast the consumed energy in order to enhance the data accuracy and the performance of the dynamic forecasting algorithms, by the implementation of the necessary algorithms responsible for the energy forecasting between providers and consumers in the power-grid of under developed countries [3, 5–7].

The implementation of forecasting algorithms for the remote prediction of load and PV energy generation in the smart grid is investigated in many recent studies.

Renato William et al. [14, 15] proposed their approach based on the integration of wireless sensor networks (WSNs) – used in the communication system with an electrical energy-measurement structure. They checked the feasibility of the largescale installation of the intelligent electronic meters in low-voltage consumer units. The disadvantages of this method are the loss of data packages during the transmission of messages and the long response time of the communications between the nodes. Krishna Paramathma et al. [17] have presented an effective methodology for the control and the optimization of energy consumption. The used method has some drawbacks since it has been demonstrated in a small-scale system. This method is modeled in MATLAB to present the energy exchange between consumers and providers; the inconvenient of this work is the high cost and the requirement of a huge response-time in the demonstration (Table 1).

Table. 1. Summarizing table of the existing research studies dealing with the smart metering

The deployment of Wireless Sensor Networks based smart grid for the control and the monitoring of the smart meters [15]	The use of Wireless Sensor Network in the measurement structure to test the feasibility of the integration of smart meters in low-voltage consumer-units	The improvement of the performance of digital electronic meters by checking the data transmission in Smart Grid and by controlling the message transmission times
Development and implementation of AMI for efficient energy use in SG environment [17]	The use of AMI to acquire the metered data of Voltage, current, power, and power factor to develop an intelligent and a cost-effective AMI infrastructure	The implementation of AMI for the smart and the cost effective metering framework that is able to acquire the energy needs from the demand side to reduce the power consumption

The contribution of this work is focused on the prediction of real-time algorithms used RBF training methods based neuro-fuzzy system for the forecasting of load in function of the input parameters as well as, the evaluation of the forecasted load data accuracy by the test of the performance of the used algorithms via the MAPE calculation. The proposed dynamic prediction algorithms is aimed to compare the predicted data versus the forecasted ones to evaluate the reliability and the accuracy of the predicted data and the performance of the developed dynamic prediction algorithms. This paper is organized as follows: Sect. 2 presents the ANN model used for the energy forecasting. Section 3 describes used load forecasting methods. The last section presents and discusses the results of the used methods with respect to the existing research studies; the conclusion summarizes the contribution of this work and discusses some perspectives dealing with demand response programs for the demand side management.

2   The Conceptual Study of the ANN

2.1   The ANN Model

The ANN have shown strong capabilities in multipurpose approaches like recognition, classification, clustering and prediction, there are the most used technology according to [21, 22] that proved the best results in the prediction compared to the newest methods such as linear, nonlinear and fuzzy logic. ANN architecture is composed generally of three layers linked together by the interconnected nodes that integrate the network as shown in Fig. 1. The first layer is used to receive input data to enter them into the model, the second layer known by the hidden layer is composed of a number of output parameters that are transferred from the first layer to be combinated and trained in the second layer, the 3 rd layer presents the obtained output from the processing and the transfer of final values in the hidden layer (Table 2).

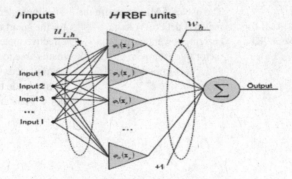

Fig. 1. RBF based ANN model [21]

3 The Proposed Approaches

3.1 The Used Methods for the Dynamic Prediction Approach

Table 2. Summarizing table of the RBFNN methods for the data learning and performance test of the dynamic prediction algorithms

	Concept	Advantages	Drawbacks
Support Vector Regression SVR [21]	Algorithms based on the fusion of groups of data according to the nearest Euclidean distance to build the support vectors based on the weight input and the locations of RBF using adjustment parameters inputs	high analysis time	poor prediction results
Extreme Learning Machine ELM [21]	Algorithms based on the use of orthogonal least squares to find the output weights and the locations of the RBF centers	Produced reasonable results Faster than other algorithms despite the use of RBF and SVR inputs	The use of both RBF and SVR inputs complicate the training
Improved Second Order ISO [20]	Iteration algorithms based on the minimization of errors expressed in 1st order by $W_k + 1 = W_K - a\,g$ and in 2nd order by the identity matrix Adjustment of all parameters related to RBF	Generates results with 10 times more precision than SVR	Execution time is very long
Error Correction ErrCor [21]	characterized by a single analysis to obtain the optimal solution	Error validation is more accurate compared to ELM and SVR Error correction execution time is faster compared to SVR and ELM	

3.2   The Dynamic Prediction Approach

The proposed dynamic prediction approach has used the combined ISO and ErrCor methods ISO-ErrCor based on the trained RBF neuronal network for the load and PV generation forecasting in short and long term related to short-term measurements of the load and PV generation, the solar radiation data as well as, the period of time (summer, holiday etc.). The proposed dynamic prediction approach is inspired from the solar radiation prediction approach [18], which was used the artificial neuronal networks and the previous short- and long-term weather information to forecast the solar radiations. It was used the forecasted solar radiation data and the previous electrical measurements from the Electrical distribution French company, to forecast the next load use and the photovoltaic generation (Fig. 2).

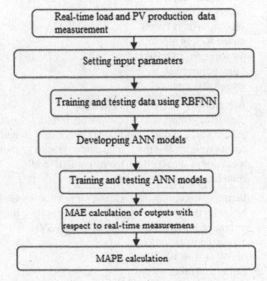

Fig. 2. ANN dynamic prediction model

3.3   The Algorithm of the Short-Term Load Data Forecasting

The Fig. 3 presented load data-forecasting algorithms based on the input parameters (previous short-term load measurements from the electrical French company, the temperature and the day type. The presented flow chart in Fig. 3 follows the real-time electrical load forecasting with training and validation algorithms based on the RBF methods used ANN in machine learning. If the forecasted load value is superior than 0, then the algorithm will be validated and the MAPE will be calculated, else the algorithm returned to the prediction process.

The Load forecast performs a short-term load forecast using a pre-trained neural network or bagged regression tree model. This function is used for the short-term load forecasting using ANN or bagged regression trees model.

function y = loadForecast (date, temperature, isHoliday)

y = loadForecast(model, date, hour, temperature, isWorkingDay)

This function shows the predictors generation as a matrix of predictor for the load forecasting model.

function [X, dates, labels] = genPredictors(data, term, holidays)

[X, dates, labels] = genPredictors(data, term, holidays)

The input parameters used to predictor generation are:

data: A Dataset array of historical weather and load information term: The horizon of the forecast.
holidays: A vector of holidays. If this is not specified, holidays are generated by the function createHolidayDates.

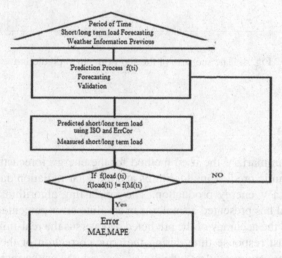

Fig. 3. Algorithm of the load consumption data forecasting

4 Results and Discussion

The dynamic prediction is used for the forecasting, training and validation of the short or long term load and PV power production. This section presents and discusses the results of the used dynamic prediction methods.

4.1 Dynamic Prediction Results

The forecasted data were used the previous measurements in «Eco-mix website» available in French electrical company and presented the scenarios of short-term load and PV power generation. The dynamic prediction algorithms were implemented in MATLAB tool, using the input parameters as the previous measurements, date type and weather information to forecast load and the PV power generation.

4.1.1 The Scenarios of the Dynamic Prediction of Load Needs

Figure 4 presents the forecasted electrical load and shows a peak power at 12 AM, as well as power drop at 5 AM. This variation is explained by the peak power, related to the electronic and domestic equipment over-use at 12 AM, added to standby state of the PCs functioning and the industrial materials. The power drop at 5 PM is related to the energy use reduction explained by period of time (break-time, holidays, weekdays...).

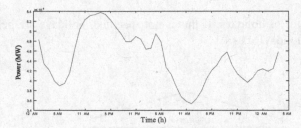

Fig. 4. The scenario of the short-time load prediction

5 Conclusion

This paper has summarized the used method for the energy forecasting. This method has used the dynamic prediction algorithms and the solar radiation data to forecast the load use and the PV energy production. The forecasting algorithms of the dynamic prediction method has presented a small mean absolute error percentage, demonstrated the reliability and the accuracy of the predicted data versus the real-time measured data, as well as, the fast response time during the error correction of the exchanged data between consumers and providers, which were considered as the main contribution of this work. This work can be extended to deal with the demand side management by using demand response programs in the context of enhancing the quality of the supply and balancing the offers and the demands.

References

1. Dönük, A., et.al.: Metering and Data Processing in a Micro-Scale Area for Smart Grid Applications. In: IEEE International Istanbul Smart Grid Congress and Fair (ICSG) (2016)

2. Bikmetov, R., et al.: Dynamic prediction capabilities of smart metering infrastructure. In: IEEE North American Power Symposium (NAPS) (2015)
3. Kukuča, P., et al.: From Smart Metering to Smart Grid. Meas. Sci. Rev. **16**(3) (2016)
4. Abdul Khadar, A., et.al.: Research Advancements Towards in Existing Smart Metering over Smart Grid. Int. J. Adv. Comput. Sci. Appl. **8**(5) (2017)
5. Zheng, J., Gao, D., Lin, L.: Smart meters in smart grid: an overview. In: Presented at the 2013 IEEE Green Technologies Conference, Denver, CO (2013)
6. K. Budka and et. al.: Communication Networks for Smart Grids: Making Smart GridReal (2014)
7. Xi, F., Satyajayant, M., et al.: Smart grid – the new and improved power grid: a survey. Commun. Surv. Tutorial **14**, 69 (2012)
8. Khan, M.F., Jain, A., Arunachalam, V., Pavethan, A.: Communication technologies for smart metering infrastructure. In: Presented at the 2014 IEEE Students' Conference, Bhopal (2014)
9. Shao, S., et al.: Smart grid distribution and consumption: communication network architecture. J. Commun. **8**, 480–489 (2013)
10. Bikmetov, R., et al.: Dynamic prediction capabilities of smart metering infrastructure. In: 2015 North American Power Symposium Conference Paper, Oct (2015)
11. Kanchev, H., Colas, F.: Emission reduction and economical optimization of an urban micro-grid operation including dispatched PV-based active generators. IEEE Trans. Sust. Energy **5**(4) (2014)
12. Eco-mix websites of Electrical Transport Network: https://rtefrance.com/fr/eco2mix/eco2mixconsommation
13. Hernandez, L., et al.: A survey on electric power demand forecasting: future trends in smart grids, micro-grids and smart buildings. IEEE Commun SurveyTutorial. **16**(3), 1460–1465 (2014)
14. Fecil'ak, P., et al.: A Non-intrusive Smart Metering System: Analytics and Simulation of Power Consumption. John Wiley & Sons (2019)
15. Renato William R. de Souza et al. Deploying Wireless sensor networks–based smart grid for Smart meters monitoring and control, John Wiley & Sons, 2015
16. Salimuddin, M., et al.: Smart metering for smart power consumption. Int. J. Res. Appl. Sci. Eng. Technol. (2019)
17. Marimuthu, K.P., et.al.: Development and Implementation of Advanced Metering Infrastructure for Efficient Energy Utilization in Smart Grid Environment. John Wiley & Sons (2017)
18. Yan X., et.al.: Solar radiation forecasting causing Artificial Neuronal Network for local power reserve. In: CISTEM (2014)
19. Prez, G., et.al.: Impact of the power consumption in the buildings in Central School of Lille, Energy Build. **123**, 8–16 (2018)
20. Xie, T.: Fast and efficient second-order method for training radial basis function networks. IEEE Trans. Neural Netw. Learn. Syst. **23**(4) (2012)
21. Bodgan, M.: A novel RBF training algorithm for short-term electric load forecasting and comparative studies IEEE Trans. Ind. Electron. **62**, 6519–6529 (2015)
22. Yadav, A.K.: Solar radiation prediction using Artificial Neural Network techniques: A review. Renew. Sust. Energy Rev. **33**(3), 772–781 (2013)

KGChain: A Blockchain-Based Approach to Secure the Knowledge Graph Completion

Ala Djeddai(✉) (iD)

LAMIS Laboratory, Larbi Tebessi University, Tebessa, Algeria
ala.djeddai@univ-tebessa.dz

Abstract. In the last recent decade, the knowledge graph become essential in many artificial intelligent applications such as link prediction, recommendations, entity resolution…etc. Knowledge graph completion aims at predicting missing triples where the embedding methods take the lion's share. These methods embed entities and relations into continuous vector spaces and use scoring functions to compute the plausibility of triples. The knowledge graph privacy and security take an important role to protect the data and the prediction model. We propose *KGChain*, a new framework which combines an off-chain storage with the blockchain in both the embedding and the completion tasks. Our work has two steps: protecting the knowledge graph and the model privacy using Fabric Hyperledger and securing the completion tasks by smart contracts. The proposed framework is evaluated with several datasets using translation-based embedding.

Keywords: Knowledge graph · Blockchain · Fabric Hyperledger · Knowledge graph embeddings · Knowledge graph integrity

1 Introduction

In recent years, a massive amount of data is published on the web due to the increasing usage of applications. Therefore, processing these data has been the goal study of many researchers where the main objective is trying to learn more about users and its behaviors. Many artificial intelligence applications use KG processing in order to help in recommendations and predicting new knowledge. Knowledge graph (KG) which connects real world entities by relations, plays a central role when the application needs to represent and reason about knowledge. One of the most important problems that face KG is the incompleteness and therefore the KG completion tries to predict new facts. Most recent approaches use the KG embeddings in order to learn low dimension vector spaces about entities and relations and use these vectors in the prediction process.

The KG security and privacy is a big challenge due to its critical content such that in social networks and healthcare. Blockchain, which is a distributed and secure database, has been used to tackle the security and privacy of KG [1, 2]. It has been recently used to keep the privacy of the personal data [3, 4] which can be represented by KG.

In this paper, we propose *KGChain* which focuses on the security and privacy problem in the KG completion using blockchain and especially when the KG embeddings

© Springer Nature Switzerland AG 2022
R. Chbeir et al. (Eds.): MIKE 2021, LNAI 13119, pp. 218–224, 2022.
https://doi.org/10.1007/978-3-031-21517-9_22

is used. To the best of our knowledge, this is the first work that uses this technology to make the KG completion more secure, trust and to keep its privacy.

The main contributions of this work are listed below:

1. Using a permissioned blockchain for keeping the integrity of the learned embedding model and sensitive information used to build and rebuild the original KG. Thus, these data are saved in a secure and decentralized manner.
2. Improving the *KGChain* scalability by using an off-chain storage in order to save only critical data in the blockchain and thus to avoid additional transactions.
3. Securing and improving the trust of KG completion tasks such as link predication and triple classification using the blockchain peers in a decentralized manner.
4. Implementing and evaluating *KGChain* by populated datasets and Fabric Hyperledger powered by Go smart contracts and using MangoDB for off-chain storage.

This paper is structured as follows: Sect. 2 presents the background and related works. Section 3 describes how securing and keeping the privacy of KG completion using blockchain and off-chain. Section 4 presents the implementation and the evaluation results with security analysis. Section 5 concludes the paper with future directions.

2 Background and Related Works

2.1 Knowledge Graph and Traditional Embeddings Approaches

Knowledge graph describes real word facts by connecting entities with relations, for example, "Algeria location north-Africa", where "Algeria" and "north-Africa" is respectively a subject and an object entities and "location" is a relation. Therefore, every KG is a set of triples and it is defined as a couple of (E, R) where E is the set of entities and R is the set of relations. The KG appeared with the Google KG in 2012, after that many KG was created and published like: Freebase [5] and WordNet [6]. It is constructed using different methods such as knowledge extraction and fusion. It has a widespread usage in artificial intelligence, such as link prediction and recommendations.

The two main components in the KG completion based on KG embeddings are link prediction and triple classification. The former tries to predict missing triples, whereat the latter classifies if a given one is true or not. Many approaches have been proposed such as translation-based models TransE [7] and its extensions. The KG embeddings is learned by different optimizers like Adam [8] by splitting the data into three separated sets: train, test and validation. We refer the reader to surveys of [9] and [10] for more information about of KG embeddings and link prediction.

2.2 BlockChain and Knowledge Graph Privacy

The blockchain is a distributed database that records all transactions executed by its distributed peers. These transactions are saved in interconnected blocks and they cannot be modified, therefore the blockchain keep the data integrity, confidentiality and privacy. This technology was born by the Bitcoin [11] peer-to-peer cash system. By the evolution and the success of the blockchain, many application domains such as the Internet of Things, Smart Grids, Knowledge management...etc., have integrated this technology in order to keep their data integrity and privacy.

In recent years knowledge graphs and blockchain have been merged to support artificial intelligence applications in many domains where the data knowledge and security play a central role. The former can be used to enhance the querying and reasoning capabilities of the blockchain where the latter is used mostly to keep the KG privacy, trust and security. The merging of these two technologies has been supported in recent years in many domains like data sharing [12] and recommendation systems [13].

The only survey that we found about privacy in KG completion is [2] where the authors investigate privacy problems in KG and propose possible solutions to protect the KG privacy under isolated setting. The authors of [1] propose a new schema to improve the KG security using the blockchain and a distributed storage system. After processing the KG files using the distributed storage, their hashes are saved in the blockchain for further verification of data integrity. In [13] a new approach of deep recommendation system using KG is proposed. The construction process of KG is different from the traditional methods by using decentralization assured by blockchain and smart contracts. The work of [12] proposes OpenKG Chain which is a network based on blockchain to share knowledge graphs in a secure and trusted manner.

3   Proposed Design

Our goal is keeping the security and privacy of the KG completion using the blockchain. To achieve our goal, we propose the architecture design which is illustrated by the Fig. 1 where our framework has five main components organized in two sides.

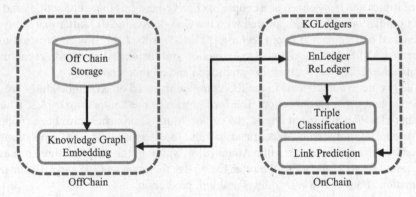

Fig. 1. Blockchain based knowledge graph completion

KG Storage: In our design, the blockchain has not been used to store the KG, but its main goal is associating the entities with its emplacements in the off-chain. All triples are stored according to their subjects and therefore, every entity is associated with its triple set where it acts as subjects. In other words, we store the outgoing edges of every entity. The JSON key value is used to describe every triple set where the object ID of everyone is associated with the subject. The triples are saved using object identifiers in order to hide their contents. The JSON description is given by the Fig. 2 where everyone has two keys: the object ID and the triple set that has a value an array of triples.

```
{"objectId": "","triples" :[{"subjectID": "", "relationID": "","objectID": ""}]}
```

Fig. 2. JSON object structure in the off-chain storage

Knowledge Graph Embedding: It acts as traditional approaches, but the KG is used in a secure and trust manner. Firstly, it interacts with the off-chain and requests all JSON objects. Secondly, in order to rebuild the KG, it requests the ledgers to resolve the associations between the object IDs and the real entities and relations. Since a permissioned blockchain is used, an authentication process is needed. For every entity, its associated hash is used to verify its triple set integrity. Finally, the embeddings can be generated and tested by splitting the KG into train, validation and test datasets. The vectors are saved in the blockchain for future utilization by the KG completion tasks.

Knowledge Graph Distributed Ledgers: we propose to use 2 ledgers, one for the entities and one for the relations where the goal is to associate the off-chain with its real entities and relations and thus keeping the KG privacy. In addition, avoiding the illegal changes and keeping the data integrity. Every asset in the ledgers contains data about entities and relations such as IDs, object IDs, embeddings and hashes. The JSON key value is used where every entry in "EnLedger" has four values and everyone in "ReLedger" has three. The entity assets contain the hashes of its triple set, in order to verify the off-chain data integrity. All components interacted with the ledgers by smart contracts. The Fig. 3(a) and (b) show the JSON of every asset in every ledger.

```
{"entityId": "","objectId": "","vecEmb": "","hash": ""}     {"relId": "","vecEmb": ""}
```
      (a) EnLedger Entry   (b) ReLedger Entry

Fig. 3. The JSON object structure of Distributed Ledgers Entries

Link Prediction: it uses the embeddings generated to predict the missed subject or the object of a relation. It uses a function to compute the scores using all entities. We propose to use it by smart contract. Thus, the process is provided by the blockchain nodes and not by one node, in order to keep the trust and the integrity of the predicted links.

Triple Classification: A given triple is classified true or false using a scoring function and a threshold. The process interacted with the ledgers to get the embeddings. It follows the same strategy as the prediction and it completed in decentralized classification.

4   Implementation and Evaluation

KGChain is implemented under Eclipse using several Java API such as: JGraphT, JSON and Fabric SDK. The Fabric Hyperledger is used with the configuration of two organizations and one peer node for each one. The fabric network uses CouchDB as a world state database and one ordering service. It was built with one certificate authority for each organization. Two channels are created for entities and relations where two smart contracts are deployed using Golang (one for each channel). MangoDB is used for the off-chain storage. For the KG embeddings, we use TorchKGE [14] which is a python module that uses PyTorch to implement several embedding models like TransE.

4.1   Experiment Configurations and Results

To test the performance of *KGChain*, a number of experiments have been performed on a machine with an Intel Core i7 processor running with a 1.8 GHz clock speed, 16 GB memory, 128 GB SSD and 1 TB storage. The components of the fabric network are deployed as Docker 2.3 images. For a best execution of TorchKGE, Python 3.8 is used with GeForce MX150 (2GB) and CUDA 10. TranE is selected with embedding dimension $= 50$, learning rate $= 0.005$, epochs $= 1000$, batch size $= 32768$ and margin $= 3.5$.

KGChain is evaluated using several datasets supported KG completion. Table 1 present descriptions about these datasets whereat Fig. 4(b) illustrates their sizes after the protection. It is observed that the off-chain sizes are extremely related to the number of triples. The ledger sizes have strong relation with the number of entities and relations.

Table 1. KG datasets used to evaluate *KGChain*

Dataset	FB15K-237	WN18RR	FB15K	WN18
Size (triples)	272115	86835	483142	141442
Entities	14505	40559	14951	40943
Relations	237	11.00	1345	18

Figure 4(a) shows the execution times of the main *KGChain* procedures. The encryption of entities and the hashing are related respectively to the number of entities and triples. The off-chain times are less than the times needed to blockchain uploading because the latter stores the data with every peer. All the previous process can be done offline but the completion tasks must be done online. The time needed for the classification is less than the prediction because the former uses only the embeddings of the triple components whereat the latter must use many entities to check the relation degree.

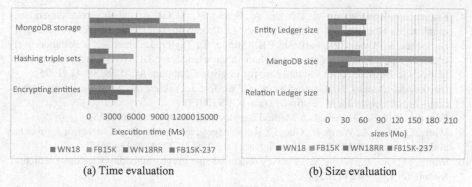

(a) Time evaluation   (b) Size evaluation

Fig. 4. *KGChain* evaluation of sizes and execution times

4.2 Security Analysis

We assume that an external adversary tries to make KG modifications or change the KG completion tasks by modifying the scoring function or the embeddings. In order to achieve this goal, the attacker must have the secret and the public keys which have been used respectively to encrypt and decrypt the entity and relation names, the hashing strategy, the certificate to access the blockchain and the smart contact, the username and password to access the off-chain. Therefore, it is very difficult to obtain all these data together. Any changing in the KG can be detected using the hash of triple sets.

5 Conclusion

In this paper, we proposed *KGChain*, which is a new approach for securing and keeping the privacy of KG completion tasks. *KGChain* has an on-chain and offchain storage using respectively Fabric Hyperledger and MangoDB. The completion tasks are performed in decentralization in order to improve their trust. The evaluation results show that it is possible to achieve the goal of securing the KG completion and keeping the privacy of the KG and the learned model. *KGChain* can be seen as a general framework which can be used to enhance the security and privacy of KG in artificial intelligence application. Future works may include extending *KGChain* to use new KG tasks such that entity resolution, entity classification and relation extraction.

References

1. Wang, Y., Yin, X., Zhu, H., Hei, X.: A blockchain based distributed storage system for knowledge graph security. In: Sun, X., Wang, J., Bertino, E. (eds.) ICAIS 2020. LNCS, vol. 12240, pp. 318–327. Springer, Cham (2020). https://doi.org/10.1007/978-3-030-57881-7_29
2. Chen, C., Cui, J., Liu, G., Wu, J., Wang, L.: Survey and Open Problems in Privacy Preserving Knowledge Graph: Merging, Query, Representation, Completion and Applications. ArXiv, abs/2011.10180 (2020)
3. Zyskind, G., Nathan, O., Pentland, A.: Decentralizing privacy: using blockchain to protect personal data. In: 2015 IEEE Security and Privacy Workshops, pp. 180–184 (2015)

4. Khemaissia, R., Derdour, M., Djeddai, A., Ferrag, M.: SDGchain: when service dependency graph meets blockchain to enhance privacy. In: Proceedings of the ACM IWSPA (2021)
5. Bollacker, K., Evans, C., Paritosh, P.K., Sturge, T., Taylor, J.: Freebase: a collaboratively created graph database for structuring human knowledge. In: SIGMOD Conference (2008)
6. Miller, G.: WordNet: a lexical database for English. Commun. ACM **38**, 39–41 (1995)
7. Bordes, A., Usunier, N., García-Durán, A., Weston, J., Yakhnenko, O.: Translating Embeddings for Modeling Multi-relational Data. NIPS (2013)
8. Kingma, D.P., Ba, J.: Adam: A Method for Stochastic Optimization (2015)
9. Wang, Q., Mao, Z., Wang, B., Guo, L.: Knowledge graph embedding: a survey of approaches and applications. IEEE Trans. Knowl. Data Eng. **19**(12), 2724–2743 (2017)
10. Wang, M., Qiu, L., Wang, X.: A survey on knowledge graph embeddings for link prediction. Symmetry **13**(3), 485 (2021)
11. Nakamoto, S.: Bitcoin: A Peer-to-Peer Electronic Cash System (2009)
12. Chen, H., et al.: OpenKG chain: a blockchain infrastructure for open knowledge graphs. Data Intell. **3**(2), 205–227 (2021)
13. Wang, S., Huang, C., Li, J., Yuan, Y., Wang, F.: Decentralized construction of knowledge graphs for deep recommender systems based on blockchain-powered smart contracts. IEEE Access **7**, 136951–136961 (2019)
14. Boschin, A.: TorchKGE: Knowledge Graph Embedding in Python and PyTorch. ArXiv, abs/2009.02963 (2020)

Enhanced Group Key Distribution Protocol for Intra Group and Inter Group Communication Using Access Control Polynomial

M. Ragunathan[1], T. Kathirvalavakumar[2(✉)], and Rajendra Prasath[3]

[1] Department of Information Technology, V.H.N. Senthikumara Nadar College, Virudhunagar, Tamil Nadu 626001, India
[2] Research Centre in Computer Science, V.H.N. Senthikumara Nadar College, Virudhunagar, Tamil Nadu 626001, India
kathirvalavakumar@yahoo.com
[3] Department of Computer Science and Engineering, Indian Institute of Information Technology, Sri City, Chittoor, Andhra Pradesh, India
rajendra.prasath@iiits.in

Abstract. In today's Internet world, group communications have become very crucial for several applications. It is essential to maintain confidentiality during communication hence it is very important to efficiently and securely distribute the common keys to the group members and target group members for encrypting and decrypting the message. This paper proposes an access control polynomial based on Chinese remainder theorem (CRT) for group key distribution (ACPGKD). Also proposes an authentication protocol for dynamic members to join or leave the group using the polynomial to keep backward and forward secrecy in inter-group and intra-group communications. It has been shown that the proposed work is secure and computationally efficient.

Keywords: Group key distribution · Rekeying · Secure group communication · Polynomial based key communication · Chinese remainder theorem

1 Introduction

Group communication plays a vital role in pay-per-use like Pay-television, online classes, OTT, online game, video conferencing and video broadcasting. A group can be a fixed one or as a varying one. In a dynamic multicast communication, members can join or leave the group at any time. In a group, all group members require a common group key namely intra-group key to communicate with each other. Outside group members called as target group members need an inter-group key when they want to communicate with one of the group members. To achieve secure group communication, the group key (GK) must be shared only to the group members. The group key management system is divided into three categories namely centralized, decentralized and distributed. In centralized group

© Springer Nature Switzerland AG 2022
R. Chbeir et al. (Eds.): MIKE 2021, LNAI 13119, pp. 225–232, 2022.
https://doi.org/10.1007/978-3-031-21517-9_23

key management group controllers (GC) are responsible for generating and distributing the common group key among the authorized group members. In a decentralized group key management scheme, larger groups are divided into sub groups, and common group key is generated and distributed by corresponding sub group controllers. In a distributed group key management scheme group members have to cooperate with each other and contribute equally to generate the group key.

The group key management schemes must fulfill the basic security requirements namely forward and backward secrecy in the group. Group controller is responsible to prevent newly joining members from having access the previously communicated data to provide backward secrecy. To provide forward secrecy, the group controller is responsible to prevent the leaving members from further accessing future communications. Whenever group members leave the group or new members enter into the group, group controller has to update and distribute the group key, which is called as rekeying.

Vinod kumar et al. [9] have proposed centralized group key distribution protocol (CGKD) based on RSA public key cryptosystem and implemented on key star and cluster tree structure. Key star structure mainly focuses on the key update phase to minimize the computation and storage load of the key server. Clustered tree based architecture has achieved scalability and minimizes the computation complexity of the key update phase whenever the members want to join or leave the group in batches. Huo Guo et al. [3] have proposed self-healing group key distribution protocol in wireless sensor networks for secure IoT communications. They have used access control self-healing group key distribution (AP-SGKD). It satisfies basic security properties with optimal storage requirement and optimal session key recovery time. The proposed protocol holds mt-wise forward secrecy, wise backward secrecy and mt revocation capability. Additionally the storage requirement of the protocol is constant and this protocol is suitable for the Zigbee network.

Alphonese et al. [1] have proposed scalable and secure group key agreement for wireless ad-hoc networks by extending the RSA scheme. It uses linear time to generate group key agreement computations, and create private keys. The partial and group signatures are dynamic. The security of framework completely depends on the RSA scheme to achieve shared key authentication and user authentication. On receipt of a group message, every member of the group can calculate their group key by using their partial signature and private key. Each member requires internal memory to store their private key, even though they belong to many number of groups. Partial signatures of the groups and public keys of the members are stored in the memory which is accessed by members and non-members of the group. Sirui et al. [7] have proposed a secure communication system in self-organizing networks by lightweight group key generation. It is based on the difference of quantization results at one device from different channels (DORCE). It uses adaptive quantizer to generate pairwise keys. The users share the group key via the difference between pairwise keys. The users of the self organizing network can flexibly join and exit, compared to chain topology and star topology.

Yanji Piao et al. [10] have proposed polynomial based key management for secure intra group and inter group communication. The proposed group key management scheme uses two kinds of polynomials. One to derive the intra group key and the other to create the inter group key. In this proposed scheme, group members and group controllers can share the intra-group key without any encryption/decryption. It reduces the number of rekeying messages during group changes. The proposed scheme drastically reduces the amount of broadcast traffic in the inter group communication. Shaukat Ali et al. [6] have proposed a scalable group key management protocol. The significance of the protocol is reducing the number of rounds in the key generation process irrespective of the group size. Also each user needs only two transmissions for the entire process of the group key generation, and the communication complexity is distributed among the group users. This protocol does not require any secure channel or trusted server for the key management. Velumadhava et al. [8] have proposed hierarchical group key management for secure data sharing in a cloud based environment. This protocol reduces the complexity whenever a member leaves the group. This system uses inverse values to update the key during leave operation. The group members can calculate the key by themselves using the inverse value.

In the literatures most of the authors generate inter and intra group keys separately for group communication. Group controller is responsible for securely sending the group key. When members of the group are dynamic there exist computational overhead to the group controller. This paper proposes a protocol for group key distribution for communicating within a group and between groups with a new access control polynomial proposed with the base of Chinese remainder theorem. In the proposed method group controller is not sending group key but user finds the key from the polynomial given by the group controller. The proposed method needs minimum processing time. Section 2 has proposed group key distribution method. Section 3 evaluates the proposed approach in term of factorization attack, time attack, forward secrecy and backward secrecy. Section 4 discusses the experimental results. Section 5 concludes the work.

2   Access Control Polynomial Based Group Key Distribution Method

Proposed an Access control polynomial based group key distribution (ACPGKD) protocol and a rekey generation procedure for the dynamic access control in a secure inter and intra group communications. Private and public keys are generated using the Chinese remainder theorem [4].

2.1   Chinese Remainder Theorem

Let m1, m2,....mn are positive integers and are relatively prime in pairs. For any given integers b1, b2,...bn, following system of congruence equations have a unique solution.

$$GK \equiv b1 \bmod m1,$$
$$GK \equiv b2 \bmod m2,$$
$$..$$
$$GK \equiv bn \bmod mn,$$

It can be represented as

$$GK \equiv bi \bmod mi \quad \text{for } i = 1, 2, \ldots .n, \tag{1}$$

where GK is a group key, and can be computed from

$$GK = \sum_{i=0}^{n} bi\, Mi\, Yi (mod\, M) \tag{2}$$

where $M = \prod_{i=1}^{n} mi$, $Mi = M/mi$ and $MiYi \equiv 1\ (mod\ mi)$ (since $MiYi$ are multiplicative inverse).

2.2 Key Initialization Phase

Let $P = \{u1, u2, \ldots .un\}$ denote the set of group members. The Group Controller (GC) generates n positive integers $m1, m2, \ldots .mn$ which are relatively prime in pairs (i. e, $gcd(mi, mj) = 1$ for $i \neq j$ and $1 \leq i, j \leq n$) and are considered as the public keys for the group members. Any given integers $b1, b2, \ldots .bn$, which satisfies the congruence Eq. (1) are private keys for the group P. As per Chinese remainder theorem, the group key GK is calculated from Eq. (2).

GC assigns the generated keys mi to all the group members and broadcast all the mis to outside group members called as target group members. The private keys bi is sent to the corresponding member ui,, $i = 1,2,..,n$ in a secured channel and members have to keep it as secret.

GC generates the public group key (PGK) by

$$PGK = \prod_{i=1}^{n} mi \tag{3}$$

The target group members have to use the PGK for communicating with the group members.

The GC generates an Access Control Polynomial (ACP) using (4) which uses public keys of all members.

$$ACP = (x - m1)(x - m2) \ldots ..(x - mn) + PGK \tag{4}$$

The group controller broadcasts the ACP to the group members and the target group members. This ACP can be used by a member or target group member for inter and intra group communications.

2.3 Key Recovery Phase

The PGK is used for broadcasting any message to the group members by the target group members. The public group key (PGK) is computed after all the group members and target group members received the ACP. The public key of any one of the member is to be substituted in x of ACP to find the PGK.

When the group members want to communicate with themselves they need exchange key (EK) of everyone in the group. Everyone in the group can find PGK from ACP by substituting their public key instead of x. Everyone can find their exchange key by (5).

$$EKi = bi * Mi * Yi \tag{5}$$

Mi = PGK/mi.

Let Yi be the Multiplicative inverse of Mi where Mi*Yi ≡ 1 (mod mi).

bi is the private key for the group member.

All group members have to share their exchange key with all other members in their group. Now every group member has an exchange key of everyone in the group. Each group member can find the group key (GK) by Eq. (6)

$$GK = \sum_{i=0}^{n} EKi(mod\ PGK), \quad where\ i = 1, 2, ..., n \qquad (6)$$

GK is a group key generated by every group member. Using the GK, all the group members can communicate with others in the group.

2.4 Key Update Phase

Whenever new members want to join or existing members want to leave the group G, the GC needs to regenerate and distribute new ACP to all the present group members and target group members.

Member Leave Phase

Whenever a member ui wants to leave the group, the GC has to delete their corresponding mi and bi from the active list and has to update the value of public group key(PGK) using Eq. (7).

$$New\ PGK = current\ PGK/mi \qquad (7)$$

The GC has to update the group key by

$$New\ GK \equiv bi\ mod\ mi, i = 1, 2, ...n - 1 \qquad (8)$$

Now GC regenerates a new polynomial

$$New\ ACP = (x - m1)(x - m2) \,.....(x - mn - 1) + new\ PGK \qquad (9)$$

All current group members can derive new PGK but relieved members cannot derive new PGK, so leaving members cannot access future communications (Forward Secrecy).

Member Join Phase

Whenever a member ui wants to join the group, the GC updates mi, bi values in the active list and update the value of public group key by

$$New\ PGK = current\ PGK * mi \qquad (10)$$

The GC updates the group key by

$$New\ GK \equiv bi\ mod\ mi, \ i = 1, 2, ..., n + 1 \qquad (11)$$

GC regenerates a new polynomial

$$New\ ACP = (x - m1)(x - m2) ... (x - mn)(x - mn + 1) + new\ PGK \qquad (12)$$

The new group member ui can derive new PGK but cannot derive old PGK. So, new members cannot derive the previous Group keys (Backward secrecy).

3   Evaluating the Proposed Approach

In this section, security strength, safety, the number of rekeying messages, storage and communication overhead of the proposed ACPGKD protocol are computed and prove that the protocol is secure against the factorization attack, Timing attack and fulfill the backward and forward secrecy requirements.

3.1   Efficiency

In the existing works of the literatures, inter and intra group keys are distinct and are send separately to the members of the group. The GC encrypts an intra-group key and sends it to the group members. GC encrypts the inter group key and sends it to target group members. But in this proposed work the group controller does not need to encrypt the key and also uses single polynomial ACP to the group members and target group members for inter and intra group communication. Normally computing the polynomial is easier than performing encryption and decryption.

3.2   Factorization Attack

In order to derive GK, the group controller sends a polynomial without any encryption. However it is not easy to guess the inter group key GK from the polynomial and it is very hard to do polynomial factorization because there is actually an $O(n\log n)$ solution to the polynomial expansion [4, 5] (n represents degree of the polynomial) and the problem for polynomial factorization is NP-hard [2].

3.3   Group Key Attack

Suppose any intruder enters into a group, the intruder has to compute the multiplicative inverse pair Mi and Yi in such a way that $Mi*Yi \equiv 1 \pmod{mi}$. The intruder has to know their bi but it can be shared only by GC to their authorized members, so their EK cannot be computed by them. To view the communication messages between intra and inter group members, the intruder should know the EK of others but it is not possible as the intruder is not in the authorized active list to get them from others. Without knowing EKs the GK cannot be computed, so the intruder cannot communicate with others in the group.

3.4   Forward and Backward Secrecy

Let u1, u2,....,un are group members and m1, m2,....,mn are the private keys generated by GC for the group members. Every group member ui received their key mi from the GC. PGK $= \prod_{i=1}^{n} m_i$ is a public group key known to the GC.

ACP $= (x - m1)(x - m2) .....(x - mn) + PGK$, is given to each member by the GC for sending PGK securely. PGK is used to encrypt or decrypt the message within a group. If a member ui wants to know the PGK, their private key mi is to be subsitituted in the ACP. Now in the ACP, only PGK is there as the first term of ACP becomes vanish.

When a member uj leaves the group, GC creates new ACP without the term (x-mj) but with new PGK as $\prod_{i=1}^{n-1} m_i$ and is sent to all the group members. If the left member uj apply their private key mj in the ACP they hold, first term of ACP becomes vanish and get the value for ACP, which is nothing but old PGK that is $\prod_{i=1}^{n} m_i$ and is not eqivalent to new PGK. So this PGK can not be used to decrypt any message of the current group members. Hence forward secrecy is maintained.

If a new member uk join the group, GC creates private key mk corresponding to the new member and is sent to the member and creates new ACP with new PGK as $\prod_{i=1}^{n+1} m_i$ and is sent to all the members currently in the group. New member uk gets PGK from the new ACP by substituting their private key mk in the ACP received from the GC. As this PGK is differed from previous PGK, new member uk could not decrypt the previous messages passed inside the group as the previous messages were created with previous PGK. Hence backward secrecy is maintained.

3.5   Re-keying Overhead

In the proposed ACPGKD scheme, when a member joins the group, the group controller needs to unicast a private key to the new member in a secure channel and multicast a new polynomial ACP to the current members in the group and target group members. When a member leaves from the group, the group controller needs to change the polynomial ACP and multicast it to the group members and target group members.

4   Experimental Results

This section shows the experimental results of the proposed ACPGKD protocol. Experiments are performed on a system with 3.3-GHz Intel Core i3–3220 processor, 4-GB RAM and the OS Windows 10. JAVA programming language with Java Runtime Engine (JRE) 1.6 is used to evaluate the performance of the Group Controller and Group Members. Here the assumption is, the group is with 3 members. The Processing time is in nano seconds (ns) for Group key generation, key recovery, Rekeying, and average communication cost. The processing times are compared with the CGKD [9] and are shown in Table 1. In the figures x-axis represents number of systems and y-axis represents consumed processing time in ns. It is observed from the table that the proposed work is better than CGKD.

Table 1. Processing time

#of systems	Generation time(ns) ACPGKD CGKD		Recovery time(ns) ACPGKD CGKD		Rekeying time(ns) ACPGKD CGKD		Avg. computing time(ns) ACPGKD CGKD	
1	133216	1034562	543456	103465	104320	1534254	573322	1201157
2	162145	1503142	702987	1503124	150298	2203780	789373	1736682
3	195136	1905482	925364	2105741	192547	3005698	1015326	2338974

5 Conclusion

The protocol for distributed group key management scheme is proposed. ACPGKD for membership authentication and rekeying for the dynamic group are proposed. The protocol provides both inter group and intra group key distribution with single polynomial. The proposed protocol has reduced the computational complexity of group controller and group members. The protocol secures against Factorization attack and guarantees the backward secrecy and forward secrecy in the group. Experimental results show that the proposed ACPGKD needs less computation time, less recovery time and lesser rekey complexity than the CGKD protocol. The same protocol may be enhanced in future to implement secure group communication in the IoT environment.

References

1. Alphonse, P.J.A., Venkatramana Reddy, Y.: Scalable and secure group key agreement for wireless ad-hoc networks by extending RSA scheme. Concurrency Comput. Pract. Experience **31**(14), e4969 (2018). https://doi.org/10.1002/cpe.4969
2. Gao, S., Hoeiji, M.V., Kaltofen, E., Shoup, V.: The Computational complexity of polynomial factorization. American institute of Mathematic, **364**, 1–5, Palo Alto, California (2006)
3. Guo, H., Zheng, Y., Li, X., Li, Z., Xia, C.: Self healing group key distribution protocol in wireless sensor networks for secure IoT communications. Future Gener. Comput. Syst. **89**, 713–721 (2018)
4. Sipser, M.: Introduction to the Theory of Computation, 2nd edn. Thomson course Technology, Boston (2006)
5. Roche, D.S., Space- and time- efficient polynomial multiplication. In: ACM International Symposium and Algebraic Computation, pp. 28–31. ACM, Seoul Republic of Korea (2009)
6. Ali, S., Islam, A.R.N., Farman, H., Jan, B., Khan, M.: A Scalable group key management protocol. Sustain. Cities Soc. **39**, 37–42 (2018)
7. Peng, S., Han, B., Wu, C., Wang, B.: A secure communication system in self-organizing networks via lightweight group key generation. IEEE Open J. Comput. Soc. **1**(1), 182–192 (2020)
8. Velumadhava Rao, R., Selvamani, K., Kanimozhi, S., Kannan, A.: Hierarchical group key management for secure data sharing in a cloud based environment. Concurrency Computat Pract Exper. **31**, e4866 (2019). https://doi.org/10.1002/cpe.4866
9. Kumar, V., Kumar, R., Pandey, S.K.: A computationally efficient centralized group key distribution protocol for secure multicast communications based upon RSA public key. J. King Saud Univ. – Comput. Inf. Sci. **32**(9), 1081–1094 (2018)
10. Piao, Y., Kim, J., Tariq, U., Hong, M.: Polynomial based key management for secure intra group and inter group communication. Comput. Math. Appl. **65**(9), 1300–1309 (2013)

Author Index

Printed in the United States
by Baker & Taylor Publisher Services

Printed in the United States
by Baker & Taylor Publisher Services